Microprocessors In Industrial Control

An Independent Learning Module from the Instrument Society of America

MICROPROCESSORS IN INDUSTRIAL CONTROL

By Robert J. Bibbero

HONEYWELL, INC.
Process Management Systems Division
Fort Washington, Pennsylvania

INSTRUMENT SOCIETY OF AMERICA

Printed in the United States of America

INSTRUMENT SOCIETY OF AMERICA
67 Alexander Drive
P.O. Box 12277
Research Triangle Park
North Carolina 27709

Library of Congress Cataloging in Publication Data

Bibbero, Robert J.
Microprocessors in industrial control.

(An Independent learning module from the Instrument Society of America)
Bibliography: p.
Includes index:
1. Process control—Data processing. 2. Microprocessors. I. Title. II. Series: Independent learning module.
TS156.8.B5. 1983 670.42'7 82-48556

ISBN 0-87664-624-0

Editorial development and book design by Monarch International, Inc.
Under the editorial direction of Paul W. Murrill, Ph.D.

Production by Publishers Creative Services, Inc.

In loving memory of my wife, Lillian, 1982.

Table of Contents

Preface

ISA's Independent Learning Modules

This is an Independent Learning Module (ILM) on *Microprocessors in Industrial Control;* it is part of the Series on Fundamental Instruments and Techniques.

Although microprocessors do not of themselves measure or control, they have become the principal tools spearheading the advance of digital techniques to all types of instrumentation and control systems. Microprocessors are small and cheap enough to be used in the most fundamental instruments, yet have sufficient power and capability to take on the job of controlling entire process units.

In order to understand and apply microprocessors, it is not necessary to know advanced mathematics, but it is necessary to look at our everyday arithmetic and logical processes with sufficient understanding to learn the new way of thinking — binary logic — that actuates digital machines. While a prior exposure to logical processes, such as relay design, would be helpful, it is not required as this module supplies all the necessary background. Computer programming exposure is also unnecessary; however it would be useful to have access to a small personal or single-board computer when applying the exercises and examples of this course.

This course is intended to teach the groundwork of digital instrumentation and the application of microprocessors to control systems engineers, process operators, and technicians. A basic understanding of electrical or electronic devices is assumed; otherwise the course is entirely self-contained. By following the steps outlined in this module, the engineer or technician with no prior digital or computer experience will acquire the ability to understand, evaluate, and choose microprocessors for specific tasks and to select and apply the most common microprocessor programming languages.

The basic thrust of the applications and examples described are from the control instrumentation field, however any technically oriented person should find this module useful in acquiring fundamental digital- and microprocessor-application skills.

Unit 1: Introduction and Overview

UNIT 1

Introduction and Overview

You are about to embark on a tour with ISA's Independent Learning Module *Microprocessors in Industrial Control*. This self-study program will introduce you to the world of digital control and to microprocessors. This is different in many respects from the analog control world with which you may be familiar. The learning module is divided into units: each unit has a set of learning objectives, an explanatory section, and questions to assure that you have reached the objectives. This first unit tells you what you must know to take the course.

Learning Objectives — When you have completed this unit you should:

A. Understand how the course is organized.

B. Know the course objectives.

C. Know how to proceed through the course.

1-1. Course Coverage

Microprocessors and microprocessor industrial-control systems are parts of the rapidly evolving field of very large-scale integrated circuit (VLSI) electronics. Since important improvements to microprocessors are still being made, it is wise to learn basic principles rather than the details of specific devices. The basic elements of digital mathematics and control are common to all microprocessor systems, but individual microprocessor types vary widely in capability and features, and there is yet little standardization. Furthermore, the applications and industries in which microprocessor control is used are quite diverse. Microprocessors can operate on a unit of information, called a "word," as small as 4 bits or as large as 32 bits, where a bit is a binary digit or number to the base 2. This is a 300 million-to-one range. The computation power of microprocessor ranges from that needed for a simple four-function calculator to the equivalent of a large main-frame computer, such as the IBM 360. Applications for microprocessors are found in many different industrial situations, such as the control of heat treating furnaces, process-control systems for large chemical plants and refineries, robot automation of machine tools and assembly processes, and general-purpose, programmable logic controllers (PLC's) replacing electrical relays and controllers in many plants.

This course will focus on the basic digital principles underlying all present microprocessors and on the application and programming methods common to all systems. It will not cover the complete range of microprocessor designs (architecture) and integrated-circuit technologies, since many of these are applied to large computers rather than control systems for industry. Examples of microprocessor operations will be given, but not the complete analysis of digital control applications, such as a commercial process controller or a robot arm. As an introductory course, the purpose of this module is to give the student the necessary background to proceed further with specific applications.

1-2. Purpose

The purpose of this ILM is to show how the principles of modern digital systems embodied in the programmable microprocessor can be applied to industrial control. To achieve this purpose, it is necessary to teach binary arithmetic and the logical methods by which the microprocessor accomplishes its elementary operations. These operations can be combined or "programmed" to achieve control objectives of any desired degree of complexity. Fundamentally, the process of applying a microprocessor to a given task requires the creation of a list of instructions, the "program," in a language that can be understood by the microprocessor. Ultimately, this language is the language of binary logic and mathematics, but there may be several stages of translation intervening between the user and the device. This course aims to teach an understanding of how the microprocessor receives and responds to instructions and to teach how to perform elementary programming tasks. This is a practical and useful skill not requiring advanced mathematical or electronic knowledge, but instead, a good knowledge of the control task that the device is to perform. Therefore, instead of being an esoteric form of knowledge, the ability to use and program microprocessors is a skill that anyone in the control and instrument field is able to learn.

1-3. Audience and Prerequisites

This ILM is designed for those who want to work on their own to achieve an introductory understanding of microprocessors applied to industrial control. The material presented will be useful to engineers, first-line supervisors, senior technicians, and maintenance personnel who are concerned with the control applications of microprocessors. The course will be helpful to students in colleges, universities, and technical schools who wish to learn the theoretical and practical aspects of microprocessors in control systems.

No formal prerequisites are required to take this course. In particular, it is not necessary to have a special knowledge of mathematics, computers, or electronic-circuit theory. It is necessary to learn a new kind of arithmetic, but this requires more a reexamination of familiar elementary principles than learning new theory. Semiconductor theory is not a prerequisite, but an exposure to electronic/electrical fundamentals familiar to most in the control instrument field is assumed. Rather than elaborate preparations, the requirements for learning digital and microprocessor concepts are an open mind and a willingness to reexamine the familiar.

1-4. Study Material

No study material is needed other than this book. As promised by the title, this is an independent, stand-alone text designed specifically for this course. Some excerpts from data sheets for specific microprocessors are included in this book. In order to design or analyze specific applications, access to the complete data sheet sets furnished by various manufacturers might be required, but this is beyond the scope of this introductory course.

Sources for additional reading and data are found in Appendix A. It is suggested also that the student study other ILMs, such as *Fundamentals of Process Control Theory*, available from ISA, since these present a broad range of material applicable to the general field of instruments and control.

1-5. Organization and Sequence

This ILM is divided into eleven separate units. The next three units (2, 3, and 4) contain the elements of digital codes and binary arithmetic needed to understand any digital system. The two subsequent units (Unit 5 and Unit 6) describe the architectural principles and the basic hardware of microprocessor systems, including essential details and analysis of a few common microprocessor device types. Units 7 through 10 explain the methods of programming microprocessors to accomplish specific tasks. These programming methods range from the most basic binary machine code to some of the "high-level" languages it is now possible to use with microprocessors and which are more nearly like natural human languages. Assembly language, BASIC and FORTH, are covered in separate units. The final unit teaches a method of developing digital-control algorithms and computer programs suitable for microprocessor controllers used in industry.

The self-study method for which this module is written permits you to study at your own pace and to complete the module in a period of time convenient for you. It allows you to adjust the speed and timing of the course to best suit your own capabilities and opportunities.

The format of each unit is consistent. First, a set of specific learning objectives is stated. It is important to carefully read and understand these objectives since the remaining material will be directed toward teaching them. Each unit will contain example problems or illustrations of specific concepts. Included in each unit there are exercises to test your understanding of the material. In Appendix C you will find the solutions to these student exercises, which you should check against your own answers.

It is recommended that you make notes or underline significant portions of your text, since it is your individual property. Ample white space and margins are provided for this purpose.

1-6. Course Objectives

When you have completed this entire ILM you should:

A. Understand the basic concepts of digital coding and why it is used.

B. Know binary arithmetic and logic as the basis for elementary microprocessor operations.

C. Recognize the basic design and architectural features of microprocessors used in control.

D. Understand the features and instruction codes of specific common microprocessors including the 6500 and similar families.

E. Know the fundamentals of programming microprocessors both in machine and high-level languages.

F. Know some of the techniques of developing digital algorithms and programs for microprocessors used in industrial control.

These overall course objectives are in addition to the objectives for each unit. The unit objectives will help you focus on the study of the particular unit.

1-7. Course Length

The basic concept of the ILM is to allow the student to proceed at his own pace, that speed which will allow him to learn best. Consequently each student will complete this course at a rate different from other students. This speed will depend on his personal experience and capabilities. Most students should be able to complete this course within a period of 20 to 30 hours.

This completes the introductory material needed to begin the study of microprocessor fundamentals. Please turn now to Unit 2.

Unit 2:
Basic Digital Concepts

UNIT 2

Basic Digital Concepts

This unit introduces the basic concepts of digital number representation and coding and the reasons for using them. These concepts are fundamental to microprocessor control and prepare you to understand digital computation and logic.

Learning Objectives — When you have completed this unit you should:

A. **Understand the distinction between digital and analog representation.**

B. **Know the reasons for using digital representation in control systems.**

C. **Understand digital coding and be able to count and read integral and fractional binary numbers and convert between binary and digital.**

2-1. Analog and Digital Control

The real world of control is a world of continuous variables. The mercury of a thermometer does not move in discrete jumps nor does the gauge needle measuring the pressure in a boiler. Above the level of discrete atomic particles and quantum physics, variables seem to change smoothly and continuously with respect to control inputs. To many people familiar with control systems, analog measurement and control seems the natural and sensible way to do things. Why bother with digital computers that can only change variables in jumps? Don't they give poorer accuracy and control than traditional analog instruments?

Actually, the opposite is true. For many control situations, the discrete numerical representation is the best way to describe and manipulate variables. Let's look at some of the chief reasons for digital (and microprocessor) control.

Stability and accuracy of control

Lower cost per function

Flexibility

Greater reliability and equipment life

Human factors favoring digital interface.

2-2. Digital Precision

A typical electronic analog controller of high quality has a drift rate of perhaps 0.1% per hour, or 5 mV in a 0-to-5-volt signal. Noise in an industrial environment adds at least another 10 mV. Thus, we must contend with fundamental limits of 0.1% to 0.3% precision, even with frequently calibrated analog instruments. If we represent process quantities by discrete numbers there is no such limitation on accuracy. If we can discriminate one unit in four of our familiar decimal numbers (that is, 1 in 9999), we are assured of a precision of nearly 0.01%. Furthermore, we can increase this precision by any amount just by adding more numbers.

As we will see shortly, it is possible to represent decimal numbers interchangeably with a sequence of simple on-off conditions, for example, an electrical circuit open or closed with a switch. Suppose, in such a case, 5 volts represents an "on" condition and 0 volts an "off" condition, with the median of 2.5 volts as the dividing line or threshold. Let the average noise level be 10 mV. Then it can be shown by statistical theory that there is almost no chance whatever that the noise will exceed 2.5 volts and so be able to confuse an "on" and "off" signal. Therefore, the original precision of 0.01% will hold for an indefinite period. This is true, even if the 5-volt signal is attenuated by being transmitted a long distance on wires, for example, because whenever the signal level falls below some desired multiple of the noise level, say 10 to 1, the original signal can be reconstituted and restored by a suitable relay or electronic circuits. (This trick, known as a "repeater" for obvious reasons, has been used since the first days of the telegraph.) Therefore, a digital signal can be sent by wires for many hundreds of miles, and even through space by satellites carrying relay repeaters, without any loss or degradation. This could not be done with an analog signal.

2-3. Versatility and Cost per Function

A modern solid-state electronic analog controller is in every respect equal to or better in performance than older pneumatic or electromechanical controllers. About 20 active components, semiconductor diodes and transistors, are required for its operation. These components now can be "printed" on a small chip of silicon, less than 0.01 inch square, with many tens of thousands of components on each chip. Thus, the equivalent of 1000 or more controllers could conceivably be placed on a single chip that might sell for $10 in large quantities. The cost-per-controller function could be lowered vastly in this way. But, in general, such special-purpose analog chips are not manufactured because of relatively low demand and the very high cost of design and tooling. This cost must be spread over many thousands of units to justify the low, mass-production price. The chips that are so designed and manufactured are general-purpose circuits that can be used for many tasks. The most general and versatile circuit that can be placed on a chip is the digital microprocessor. The microprocessor is versatile because it can be *programmed* to perform an almost unlimited variety of computing tasks. By programming, we mean that the circuitry follows a sequence of coded orders that is carried in another component called a *memory* or *store*. The processor responds to each of these coded orders as if it were reconnected to perform a new task. Since these tasks can be carried out in microseconds, they can be switched and alternated so rapidly as to give the effect of a large number of different devices, all working in parallel. In other words, the microprocessor jumps rapidly between tasks, changing its function with every jump, so that it seems to a relatively slow process to be many different devices. This versatility and multiple functionality stem from the digital nature of the microprocessor that permits it to be controlled by a rapidly changing program. The versatility is enhanced even more by letting the results of a sequence of steps alter the program sequence that follows. Thus, the microprocessor control can even adapt itself to events that occur in the process.

The microprocessor is a means of providing the function of many special devices (such as controllers) at relatively low cost, through its high speed and time-sharing ability. It also can provide all sorts of functions not usually associated with controllers, such as binary logic, sequencing, and numerical calculations.

The present cost of very large-scale integrated circuit chips (VLSI) like the general-purpose digital microprocessor and its associated memory and peripheral chips is low and continues to become lower, following an exponentially decreasing curve of cost per function that began with the invention of the transistor at Bell Laboratories in 1948. Circuit fabrication on silicon is a photographic etching process, much like an elaborate form of lithographic printing, but on a microscopic scale. The cost of a chip and its processing remains relatively constant; it follows that the component cost or cost per function decreases as the technique of printing smaller and finer components advances, that is, as more components can be fitted on a standard-sized chip.

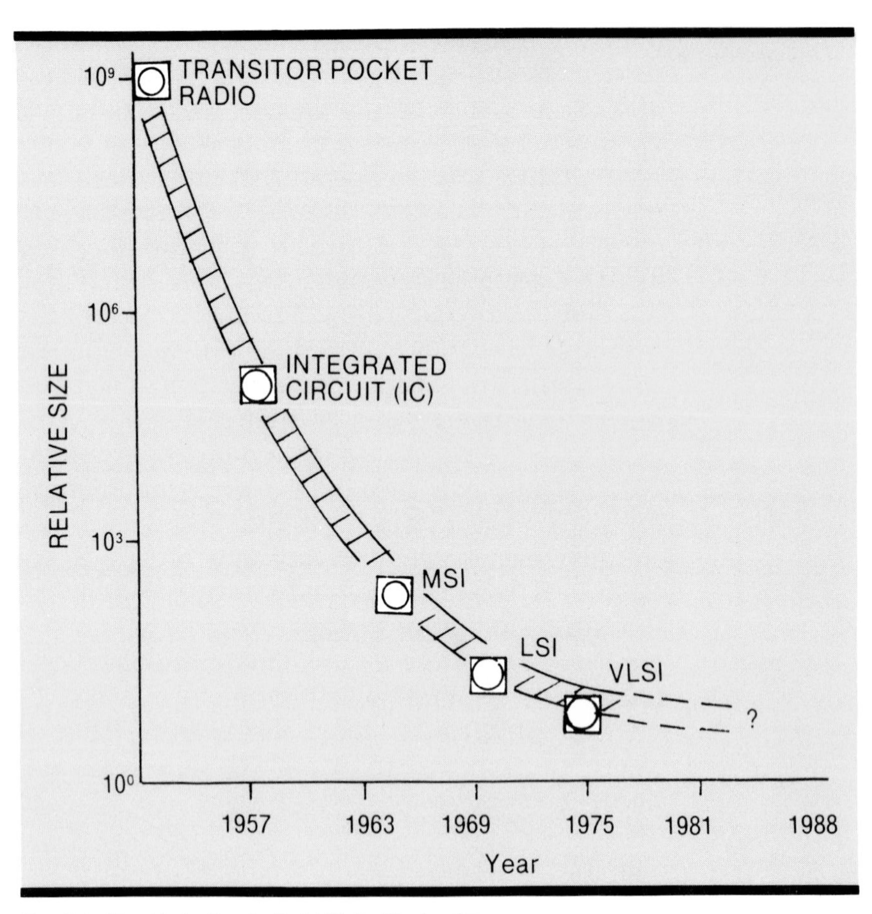

Fig. 2-1. Size Reduction in Solid-State Electronics

This technology continues to advance. As a matter of interest, the size of solid-state electronic devices decreased 10,000 fold in the six years between the introduction of the pocket transistor radio (1957) and the first integrated circuits (ICs) in 1963. This size reduction has continued at about the same rate.

It has been estimated that another size reduction on the order of 100,000 times smaller than the early ICs is possible before fundamental physical laws call a halt to the shrinking process. Clearly, the future of microprocessors is highly optimistic and justifies the effort you are expending to understand how to use them!

2-4. Reliability and Life

So we have shown that digital integrated circuit electronics is more precise and stable than analog circuitry, has lower cost per function, and can perform a far wider variety of tasks in one device. The lower cost is a result of the shrinking of tens of thousands of circuits onto a tiny, mass-produced chip. Another logical consequence of this smallness is that less power is needed to drive the circuitry, so there is less wasted power to dissipate as heat and lower voltages can be used. The result is reduced stress which leads to greater reliability and longer life. Another factor vastly improving reliability of microchip devices is the absence of external soldered connections (other than the few tens of pin connections). These are always a prime cause of failure in electronic devices. Printing thousands of components on a chip means that many tens of thousands of soldered interconnections are eliminated. The printed connections usually will not fail until the entire chip fails.

Fig. 2-2. Digital Interface with Humans

2-5. Interface with Humans and Computers

The last favorable factor for digital electronics was the improved interface with people, called "human factors." Reflection will show that for many human tasks, not necessarily including "clock-watching," numbers are a more precise and faster means of supplying information than dials and pointers. Of course, numbers are the natural output of digital devices. Furthermore, the versatile microprocessor can perform many operations on the numbers before presenting them to humans, such as converting them to decimal form or to engineering units, highlighting limits by flashing or changing color, etc. In the event that an analog output, such as a bar graph, is more appropriate to the data presented, it is easy to convert digital signals to analog ones; certainly at the low analog accuracy usually demanded by analog displays (0.5% to 1%). It is not quite so easy to convert analog signals to digits, however, this can be done by the microprocessor "with a little help from its friends," e.g., peripheral circuit devices.

Finally we must consider the system's aspect of control. Many separate control devices must be coordinated to accomplish the so-called higher level tasks such as the overall optimization of a process unit or entire plant. Today, high-level control coordination usually means a supervising computer, given hierarchal authority over individual process-control loops. And, almost always, it means a central control room or area in which plant variables are displayed on digitally driven cathode ray tube (CRT) monitors. Since the supervisory or display computers are digital, they can best communicate with lower-level digital microprocessor controllers, since both "speak the same language." Consequently, errors can be eliminated.

2-6. Digital Codes: Decimal System

To benefit from the advantages of digital "languages" we must become fluent in them. This means we will have to learn to read, write, and perform simple arithmetic in digital codes. There are many of these codes, but fortunately only a few are needed to understand microprocessors. These are the "natural" binary, octal, hexadecimal, binary-coded decimal (BCD), and one that you already know: the decimal system! We will begin by taking a fresh look at this familiar system.

Ordinary decimal numbers are written and evaluated in accordance with their "place" or the column number, counting from the decimal point. Thus the number

3421.

is written as "three thousand, four hundred, and twenty-one." Each digit to the left of the decimal point is multiplied by a power of 10 equal to the number of other digits between it and the decimal point. In "scientific notation" the above number could also have been written in terms of a series of powers of ten:

$$3\times10^3 + 4\times10^2 + 2\times10^1 + 1\times10^0$$
$$\text{or } 3000 + 400 + 20 + 1 = 3421.$$

Fig. 2-3. Intelligent Three-Toed Sloth

The exponent of 10 in each term is equal to the distance between the multiplying digit and the decimal point.

In order to generalize on this observation we will call the number being exponented the "base" or "radix" of the number system. The decimal system (from the Latin decima = 10) then consists of numbers to the base 10 or radix 10. The number of digits (or symbols) used by the system is also equal to the base. In the decimal system these symbols are the familiar Arabic numerals 0, 1, 2, . . . 9. A more general name for the "decimal point" is "radix point."

There is nothing sacred or unique about the decimal system. It is simply the consequence of the fact that mankind first learned to count on the five fingers of each hand (fingers = digits). An intelligent three-toed sloth might learn to count by threes or sixes, giving rise to another "digital" system based on sloth toes rather than human fingers.

2-7. Binary System

A digital electronic system is very reliable and precise, as noted in Sec. 2-2, if its circuits are made to operate in only two meaningful states: "on" and "off." A system having two states is called "binary" (Latin bis = two ways) or to the base or radix 2. Binary numbers are natural to the "intelligent computer" because it has only two "toes" or digits to count with: on and off. Because the base is 2, there are only two symbols used. Arbitrarily we can borrow the first two Arabic numerals, 0 and 1. These have the same meaning as in the decimal system, namely, 0 means "none are present" and 1 means "one is present." These two symbols might denote the absence of an electric current or its presence, a low and a high voltage, or almost any other conceivable two states that can be distinguished from each other. We will not be concerned now about what the two symbols represent, just that they are the only two digits in the binary system.

Just as in the decimal system, binary numbers are written in columns with each succeeding column to the left to the radix point representing an increasing power of the base 2. The binary number

$$11010.$$

is then translated as

$$1 \times 2^4 + 1 \times 2^3 + 0 \times 2^2 + 1 \times 2^1 + 0 \times 2^0$$

or in decimal numbers:

$$16 + 8 + 0(\times 4) + 2 + 0(\times 1) = 26.$$

Note that each of digits in a binary number is multiplied by a power of 2 equal to the distance (in columns) to the binary point. The value or weight of each column is thus represented by its place in a table of the powers of 2, such as:

2^{10}	2^9	2^8	2^7	2^6	2^5	2^4	2^3	2^2	2^1	2^0
1024	512	256	128	64	32	16	8	4	2	1

The column weight in a binary number is multiplied by one of the two possible digits, 0 or 1, which is equivalent to saying that the value of the binary number is the sum of the columns that correspond to the 1's. As an example, let us count the numbers equivalent to decimal 0 through 7, using three columns of binary digits:

Column number	3	2	1						
Column weight	4	2	1						
Binary number	0	0	0	=	0	+0	+0	=	0 (decimal)
	0	0	1	=	0	+0	+1	=	1
	0	1	0	=	0	+2	+0	=	2
	0	1	1	=	0	+2	+1	=	3
	1	0	0	=	4	+0	+0	=	4
	1	0	1	=	4	+0	+1	=	5
	1	1	0	=	4	+2	+0	=	6
	1	1	1	=	4	+2	+1	=	7

Note that there are eight separate numbers that can be assembled from three binary digit columns; these are the eight possible combinations of two things (0 and 1) arranged in three columns, which total up to 2 to the third power, or 8. Four columns would result in 2^4 or 16 combinations and numbers. In general, the quantity of different binary numbers that can be constructed from "n" columns of digits is 2 to the nth power. Note that a binary table such as this makes a pleasing regular pattern of 1's and 0's. This pattern will be encountered again when we consider the "truth tables" of digital logic.

We mentioned earlier that there are many possible binary codes that can represent the natural numbers. The one just explained is called the "normal binary code." Other binary codes may use redundant digits (BInary digiTs are called "bits") to aid in detecting or correcting errors and for other purposes. Such redundant bit codes are especially important when transmitting data over long distances or in an environment (such as industry) where electrical noise tending to cause errors is prevalent. We will not consider these codes at this point, but only the normal binary system.

As another alternative, a code can be constructed so that an ascending or descending number table changes by only one bit at each step. This differs from the normal code, where (as in the table above) all bits can change value in only one step, e.g., from 3 to 4 all bits change, but between 4 and 5 there is only one bit changing. Codes that are designed to vary by only one bit per step are called "reflecting codes." One of these (the Gray code) is employed in mechanical or optical transducers used to change rotary motion to binary signals. In such a transducer, misalignment of one brush or optical pickoff can cause a minimum error of but one least significant digit in the number, or 1 bit. This can be seen in the brief Gray code segment below:

GRAY CODE

0 = 000
1 = 001
2 = 011
3 = 010
4 = 110
5 = 111
6 = 101
7 = 100

Of all possible binary codes, only two are important in digital arithmetic for microprocessors. These are the normal binary and the binary decimal code, BCD. Later we will discuss the BCD code.

2-8. Fractional Numbers

So far we have considered only integers: whole numbers. These consist of digits only to the left of the radix point. In the decimal system, numbers to the right of the decimal point have the same meaning as those to the left, except that the exponent of the base, 10, is taken as negative. That is, numbers to the right are multiplied by the reciprocal of the appropriate power of 10. The decimal fraction 0.125 can be written:

$$1 \times 10^{(-1)} + 2 \times 10^{(-2)} + 5 \times 10^{(-3)}$$

$$\text{or } 1/10 + 2/1000 + 5/1000 = 0.125$$

In the same way, a binary fraction is the sum of the digits weighted (multiplied) by the negative exponents of the base 2. A short table of these weighting exponents to the right of the binary point should be written as an aid:

	$2^{(-1)}$	$2^{(-2)}$	$2^{(-3)}$	$2^{(-4)}$	$2^{(-5)}$	$2^{(-6)}$
or	1/2	1/4	1/8	1/16	1/32	1/64

Thus, the number

0.1010

is equivalent to

$$1/2 + 0 + 1/8 + 0 = 5/8 \text{ or } 0.625 \text{ (decimal)}$$

Note that the precision of a binary fraction is limited by the number of bits carried by the computer. A digital storage limited to 8 bits would be precise to $1/2^{(-8)}$ or 1/256, which is approximately 0.004 (decimal). This means that all decimal fractions cannot be expressed exactly by a limited number of bits. Given 8 bits, for example, the decimal fraction 0.500 can be represented exactly, and 0.504 can be represented to a precision of better than 0.001. However, this is not true of the decimal fraction 0.502.

2-9. Binary to Decimal Conversion

You may have noticed already that we have used an easy and straightforward method of converting binary numbers to their decimal equivalents, that is, we have multiplied each binary digit by the appropriate exponent of the base 2; the weight of each column to the left or right of the binary point. For the integers (left of the binary point) the first eight column weights are:

Bit number:	8	7	6	5	4	3	2	1
Two's exponent:	7	6	5	4	3	2	1	0
Column weight:	128	64	32	16	8	4	2	1

And for the fractions (extending the earlier table):

Two's exponent:	-1	-2	-3	-4	-5	-6	-7	-8
Column weight:	1/2	1/4	1/8	1/16	1/32	1/64	1/128	1/256

So we must merely add the weights of the binary number's columns that have 1's in them.

Example 2-1
Convert binary to decimal.

```
10010110. = 128+0+0+16+0+4+2+0 = 150
   0.1010 = 1/2  +  0  +  1/8  +  0 = 5/8 or 0.625
    11.01 = 2  +  1  +  0  +  1/4 = 3.25
```

Later, we will see that the hexadecimal system (base 16) provides a quicker and easier path for binary-decimal conversion. However, it is good to practice with the above method, at least until you memorize the weights up to at least the sixteenth power.

Converting decimal to binary integers is not much more difficult. Divide the decimal number by 2 to obtain the first (rightmost or least significant) bit, called LSB. If there is a fractional remainder (1/2) the LSB is 1, else it is 0. Divide the integers in the remainder again by 2 to obtain the next most significant digit, until there is no more integral remainder. For example:

Example 2-2
Convert decimal 26 to binary.

```
26/2 = 13 +   0 : 0 = LSB
13/2 =  6 + 1/2 : 1
 6/2 =  3 +   0 : 0
 3/2 =  1 + 1/2 : 1
 1/2 =  1 + 1/2 : 1 (MSB or most significant bit)
```

To prove this is correct, reconvert the binary result 11010. to decimal using the table of column weights:

$$16 + 8 + 0 + 2 + 0 = 26$$

The column weight table can be used directly for conversion. Subtract the largest weight that gives a remainder. Put a 1 in that column (MSB). From the remainder, subtract the next smaller weight; if it is bigger than the remainder, put down a 0 for that bit and try the next smaller weight. Continue until all column weights are used. (This is easier than it sounds!) Again, using decimal 26:

Example 2-3
Convert decimal 26 to binary.

```
 26
-16: 1  (MSB)
----
 10
 -8: 1
----
  2
 -4: 0  (too large)
 -2: 1
----
  0
 -1: 0  (LSB)
```

or, as before, 26 (decimal) = 11010 (binary).

Converting decimal fractions to binary by either method is clumsy but workable.

Example 2-4
Using the subtraction method, convert decimal 0.8125 to a binary fraction:

```
0.8125
-.500  : 1  MSB
------
.3125
-.250  : 1
------
.0625
(-.125): 0  (too large)
.0625  : 1  LSB
```

So the answer is 0.1101, which can be reconverted to decimal:

0.1101 (binary) = 1/2 + 1/4 + 0/8 + 1/16 = 0.8125 (decimal).

Exercises

2-1. *How many decimal digits are required to equal the precision of a 10-volt analog signal in the presence of 20 millivolts (mv) of noise?*

2-2. *How many binary digits are required under the same condition as question 2-1?*

2-3. *"Analog measuring devices such as pressure transducers and recorders produce a smooth, stepless output and are therefore capable of perfect accuracy." True or false?*

a) *If your answer is false, list possible sources of inaccuracy or imprecision.*

2-4. *Construct a four-column table of binary numbers. What is binary 16?*

2-5. *A computer stores numbers to a precision of 16 bits and displays five decimal-place fractions when running in the BASIC language.*

a) *A computation results in an exact decimal value of 0.1 (1/10). What is the binary number stored in the computer?*

b) *What number would be displayed by the computer (decimal)?*

c) *You have purchased a personal computer of the above specifications in your local electronics hardware store. Multiplying 0.1 by 2 results in a display of 0.19998. Should you return it to the store for repair or replacement?*

2-6. *Our system of time (60 minutes to the hour, 2×12 hours/day) comes to us from the ancient Babylonians. A Babylonian legend claims their knowledge of time came from their god "Gilgamesh, who descended from the stars." (Read, a space traveler.) Can you cite any evidence tending to support such a legend?*

2-7. *American spacemen landing on Mars found a mathematical manuscript written by early Martians. It contained the strange statements, "2 + 1 = 10" and "2×2 = 11." What number system was used by these ancient Martians?*

a) *Construct a table of "Martian" numbers equivalent to decimal 0 through 9.*

b) *If the Martians had digital computers based on their number system, what disadvantages would they have compared to our binary computers? What possible advantage?*

Unit 3: Other Digital Number Systems

Unit 3

Other Digital Number Systems

Understanding binary numbers is a prerequisite of digital computer knowledge because, ultimately, they are the only language the computer understands: the "machine code." But binary notation is very hard for humans to read without error. We will introduce two new digital notations, octal (base 8) and hexadecimal (base 16), that are easier to use and universally employed as a "shorthand" for binary and for *assembly programming*, which we will study in another unit.

In this unit we will study two other notational concepts also, negative numbers using *two's complements* and binary coded decimal, both of which must be understood before going ahead with digital arithmetic and logic.

Learning Objectives — When you have completed this unit you should:

A. Understand the octal and hexadecimal notations and why they are useful in digital computing.

B. Know how to convert numbers between binary, octal, hexadecimal, and decimal systems.

C. Understand how to represent negative numbers using the two's complement form.

D. Be familiar with binary coded decimal (BCD) notation.

3-1. Octal Notation

Understanding the octal (base or radix 8) system should not be difficult after you have completed the previous unit. Just imagine you have only eight fingers and count using just the symbols 0 through 7. These symbols in the octal system are identical to those of the familiar decimal system, but when we arrive at "8" we must call it "10," since we have reached our self-imposed limit of eight symbols. Following 8 (written "10") we continue with 11, 12, etc. as in the following table.

Decimal	Octal
1	1
2	2
3	3
-	-
-	-
7	7
8	10
9	11
10	12
-	-
-	-
15	17
16	20

Table 3-1. Octal Notation

The column weights (decimal multiplier) for each "place" to the left of the radix point are the positive powers of 8:

Column	4	3	2	1
Power	8^3	8^2	8^1	8^0
Decimal multiplier	512	64	8	1

and to the right are the negative powers: 1/8, 1/64, etc.

3-2. Octal and Binary

Many older computers used octal numbers for addresses or machine code instructions. However, the chief advantage of octal numbering is that it is much easier to use than binary, and yet it is simple to convert back and forth between the two systems. Recalling that three binary digits allow eight combinations, these can be assigned to the symbols 0 through 7. Thus, a group of three binary digits always can be interpreted as an octal number and vice versa. However, it must be clearly understood that digital computers **cannot** "read" octal numbers directly, but only the binary code. Octal (and hexadecimal) notations are only for our convenience.

For example, the binary number 100101110. looks forbidding but it can be converted to octal by merely grouping the binary digits in threes, starting at the radix point, and writing under each the octal equivalent (which is the same as the decimal equivalent, since we only use the first seven digits).

100	101	110	binary
4	5	6	octal

Since we share symbols between octal and decimal, we must be careful to avoid confusing the two, therefore if there is any chance of a mixup, we denote the base by means of a subscript or spell it out:

$$456_8 \text{ or } 456 \text{ octal}$$

The expression "456 octal" is quite a different number than 456 decimal; using the weights given above we see the octal number is equal to:

$$4\times64 + 5\times8 + 6\times1 = 302 \text{ decimal}$$

To convert octal numbers to binary, just reverse the "group by threes" procedure. Thus the octal number 237 is converted to binary as follows:

Example 3-1

2	3	7	octal number
010	011	111	binary group of threes
	or	010011111	binary number

3-3. Octal to Decimal Conversion

Octal numbers are best converted to decimal using the weight table indicated earlier. Taking another example, 237 octal is:

$$2\times64 + 3\times8 + 7\times1 = 159 \text{ decimal}$$

Converting decimal numbers to octal is very much like the equivalent conversion of decimal to binary, which, as we learned in the last unit, was accomplished by repeated division using the base 2 as a divisor. This is known as the "dibble-dabble" (or sometimes double-dabble) method and will work for any base. To obtain octal from decimal numbers, we repeatedly divide by 8 and call the remainder after each step the next octal digit, starting with the least significant (LSD) and ending with the most significant digit (MSD). As an example, let us reverse the last operation and convert decimal 159 to octal.

Example 3-2

Decimal	Remainder
159/8 = 19 +	7 (LSD)
19/8 = 2 +	3
2/8 = 0 +	2 (MSD)

Writing the result as usual from left (MSD) to right (LSD), the answer is 237 octal.

3-4. Hexadecimal

The hexadecimal notation uses the base 16. By this time you should recognize the rules of the game! Each column or place to the left of the radix point represents a multiple of 16 to a power which is one less than the column number:

Column	4	3	2	1
Power	16^3	16^2	16^1	16^0
Decimal multiplier	4096	256	16	1

The columns to the right of the radix point represent negative power multipliers as usual: 1/16, 1/256, etc. Since the base is 16, there must be 16 symbols beginning with 0. But wait a minute! Our Arabic numerals only allow us ten! We can use 0 through 9, but for the numbers equivalent to decimals 10 through 15 we must invent some new symbols. The easiest solution for the book printer and for our memory process is to use the letters A,B,C,D,E,F. Consequently, our hexadecimal table will look like this:

Decimal	Binary	Octal	Hexadecimal
0	0000	0	0
1	0001	1	1
2	0010	2	2
3	0011	3	3
4	0100	4	4
5	0101	5	5
6	0110	6	6
7	0111	7	7
8	1000	10	8
9	1001	11	9
10	1010	12	A
11	1011	13	B
12	1100	14	C
13	1101	15	D
14	1110	16	E
15	1111	17	F
16	10000	20	10

TABLE 3-2. Hexadecimal Notation

Hexadecimal ("hex") numbers are widely employed to represent addresses and operation codes in microprocessors. Each hex number represents a unit of four binary digits (bits) which is often called a *nybble* (or nibble) since it is half of a byte (8 bits). The byte is the standard unit of microprocessor memory and arithmetic since most microprocessors are constructed to operate on 8 bits at a time. Later we will see that this is because the channel to the arithmetic portion of the microprocessor, called the *data bus*, consists of eight parallel wires or their equivalent, and consequently the components (ports and registers) at the terminations of this bus are designed to accept 8 bits at a time.

A byte, therefore, consists of two hex numbers (2 nybbles or 8 bits). Eight bits allow 2^8 or 256 combinations of 0's and 1's, which is enough to differentiate between all the instructions of an 8-bit microprocessor. So instructions can be named by combinations of two hex numbers. Likewise, the number of memory locations that can be directly addressed by an 8-bit microprocessor does not exceed the number of combinations allowed by 16 bits (2^{16}), or 65,536 (as we will see in a following unit, this is because the address bus usually has 16 conductors). Consequently microprocessor addresses can be expressed by four hex digits.

Example 3-3

A typical microprocessor instruction in hex code might look like this:

LDA (Load accumulator in memory) = A5

where A5 is understood by the computer to be the eight binary-digit combination:

A (=decimal 10) = 1010 (MSD binary)
5 (=decimal 5) = 0101 (LSD binary)

or 10100101

You can see that this grouping of 4 bits maps hex digits very simply into bytes, the natural unit for 8-bit processors. But this kind of scheme would not work well with octal numbers, which represent only 3 bits. Hex numbers are so useful that they are given a special symbol, the dollar sign ($) so that a hex address of 2 bytes would be written as $A001, for example, instead of A001 sub 16, if there is any chance of confusion. The $ sign is favored because it is much easier to enter or print from a computer than the subscript, therefore we will use it to denote hex numbers henceforth in this text.

To sum up, hex-to-binary conversions are accomplished by grouping binary numbers in fours (with the aid of Table 3-3, for example) and vice versa. Each hex number represents half a byte or a nybble, and thus serves as a convenient shorthand for binary code used by the microprocessor for operation codes (op codes) and addresses (of which more later). In common operations such as *offset,* locating the address number of a given datum in memory relative to another address, we will need to do binary and hexadecimal arithmetic, especially addition and subtraction. This will require a knowledge of negative binary numbers, which we will introduce at the end of this unit.

3-5. Hexadecimal and Decimal

Often addresses of memory locations are given in decimal, especially when using higher-level computer languages such as BASIC. It is important to know how to convert back and forth between the two number systems. The methods are the same as we used with octal. The column weight table for powers of 16 given above is employed to go from hex to decimal. Merely multiply each hex digit by its column weight and add the results.

Example 3-4
Convert $A13F to a decimal number.

Column	4	3	2	1	
Hex weight	16^3	16^2	16^1	16^0	
(Decimal)	(4096)	(256)	(16)	(1)	
Hex number	A	1	3	F	
Decimal multiplier	10	1	3	15	
	10×4096	1×256	3×16	15×1	
Products	40960 +	256 +	48+	15	= 41279 decimal

Another and perhaps easier method is to first convert the hex number to binary by groups of four and then to decimal.

Example 3-5
Convert the least significant byte of A13F to decimal.

		3		F	hex			
		0011		1111	binary			
Weights	128	64	32	16	8	4	2	1
	0	0	1	1	1	1	1	1
			32+	16+	8+	4+	2+	1=63 decimal

To go the opposite way, from decimal to hex, use the "dibble-dabble" method, but divide by 16 instead of 8 or 2. Remember, the d-d method will work for any base. Remember, also, to convert your decimal remainders to hex numbers ABCDEF if they exceed 9.

Example 3-6
Convert decimal 23456 to hex.

23456/16 =	1466 + 0 = $0 (LSD)
1466/16 =	91 + 10 = A
91/16 =	5 + 11 = B
5/16 =	0 + 5 = 5 (MSD)

Answer: $5BA0

3-6. Negative Numbers

Addition and subtraction are the most common arithmetic operations performed by a digital computer. Because it is so frequently used, addition is usually carried on by "hardware," meaning a wired sequence of transistors carrying on logical operations in a fixed order, so that the programmer (you) will not have to go through all these logic steps each time you want to add two numbers. Instead, you merely command the computer to perform this operation with an easily recalled name (mnemonic) such as ADD or ADC, having a hex code such as $87 or $6D (depending on the processor) and follow this code with the addresses (locations) where the numbers to be added will be found. (In a unit to follow we will see just what these logical steps are.) Since the operations for adding are different from those for subtraction they require a new set of hardware to carry them out. In addition, we must know which of two numbers is larger and the sign (+ or −) of each in order to tell whether the result is positive or negative.

All this can be avoided if we can convert between positive and negative numbers, since if we add a + number to one that is −, the result is the same as if we had subtracted. In this case, we need only have an "adder" circuit in our hardware.

In *signed* notation, a negative number is expressed as a sign followed by a magnitude. Thus, the negative number −3.52 has two parts: the sign (−) and the magnitude (3.52). If the number was changed to positive, the sign would be written as + (or omitted and understood to be +) but the magnitude part would not change. To add two numbers in this notation we must look to see if the signs are the same or different and compare the magnitudes before we can determine the sign of the result. Since this is a nuisance and requires extra hardware, we must look for a better way. There is one, which we call the *complement* method.

Consider a mechanical counter consisting of a fixed number of wheels, say four, each carrying the ten decimal digits, and a fifth wheel or flag which carries the minus sign. When such a counter passes the boundary between + and − numbers, it will show something like this:

+0003
+0002
+0001
+0000
−0001
−0002,
etc.

If we wish to cover the entire range from +9999 to −9999 then the additional column or flag for the sign is necessary. Suppose, in order to save the added hardware, we are willing to limit the number range, say from +8999 to −1000. Then we could represent the same numbers using the following scheme or code:

+2 = 0002 (as before)
+1 = 0001
+0 = 0000
−1 = 9999
−2 = 9998
etc.
−1000 = 9000
+8999 = 8999

With this scheme we have avoided the need for a − sign, since the presence of the number 9 in the fourth column (MSD) is a signal that the number must be negative. Any other digit in the MSD column means a + number.

3-7. Complementary Notation

What have we done here? We have transformed the negative numbers by *complementing* them. The *radix complement* is what results from subtracting the magnitude of a signed number from the radix or base, 10 in this case. The *diminished complement* is obtained by subtracting from the radix minus 1, or 9 in this case. We have created a new and useful notation for negative numbers by using the radix or ten's complement form. With this form, we avoid the requirement for a sign.

To obtain the radix complement, first calculate the diminished complement and then add 1.

Example 3-7
Find the ten's complement of −2.

−2	Sign-magnitude form
0002	Magnitude only
9997	Nine's complement
+1	Add 1
9998	Ten's (radix) complement

Another way of obtaining the ten's complement of a decimal number is to subtract its magnitude from 10 to the nth power where n is the number of digit columns we are using. In the above case, subtract 2 from 10000 (10^4) to obtain the ten's complement 9998.

The radix complement method of representing numbers is much more useful and simpler, as well, when it is applied to the binary system where the computer does its arithmetic. It is very simple in terms of computer hardware as well as conceptually to obtain the diminished complement of a binary number: all you must do is substitute 1's for 0's and vice versa. So the rule for obtaining the radix (two's) complement for binary numbers is:

Convert all 1's to 0's and 0's to 1's.
Add 1 to the result.

Example 3-8
Find two's complement of binary 5 or 0101.

0101	Binary 5
1010	One's complement
+1	Add 1
1011	Two's complement

The two's complement notation is used in most microprocessors to allow representation of negative numbers in order to avoid the need to reserve a bit for the sign. As we will see in the next unit, subtraction can be carried out by two's complementing and adding. The sign is automatically accounted for in this method. Thus, we can use one hardware "adder" for both addition and subtraction (but we need a one's complementer, which is quite simple). Some larger computers use one's complements for the same purpose, but this requires additional hardware and computing time. So, for microprocessors, the usual rule is:

Express positive numbers in normal binary form.

Express negative numbers in two's complement form.

Subtract a number by adding its two's complement.

The following 4-bit table of two's complements will help fix this concept.

Positive Binary	Decimal Value	Negative Binary
0111	7	1001
0110	6	1010
0101	5	1011
0100	4	1100
0011	3	1101
0010	2	1110
0001	1	1111
0000	0	0000

Table 3-3. Two's Complements

We close this subject by noting that the two's complement form of a negative number is unambiguous as far as the computer is concerned but is difficult for us humans to read without a conversion table such as the above. By recomplementing the number resulting from two's complement arithmetic we recover the original binary form. Thus, the two's complement of −5 can be converted back to binary 5 (and a mental note made of the sign) as follows:

0101	Binary 5
1011	Two's complement
0100	One's recomplement
+1	Add 1 (two's recomplement)
0101	Binary 5 regained

3-8. Binary Coded Decimal

The idea behind BCD is very simple. Each of the decimal digits 0 through 9 is represented by a binary code. Using the normal (8-4-2-1 weights) binary, we must reserve 4 bits for each digit. Thus, BCD is identical to the hexadecimal conversion, which also uses 4 bits per symbol, except that any combination greater than 9 (1001) is invalid.

The reason for BCD is to permit decimal arithmetic in such cases as accounting, where the fixed precision of 0.01 (dollars) is demanded, regardless of the length of the computation. Ordinary binary arithmetic would not do because of the round-off errors, as we showed in the previous unit. If we add $5.00 and $5.00 it is not acceptable to display $9.999999999! Using BCD, the results of any computation are exactly as they would be using decimal units.

The BCD decimal conversion is shown in the following table:

Decimal	BCD
0	0000
1	0001
2	0010
3	0011
4	0100
5	0101
6	0110
7	0111
8	1000
9	1001

To represent a two-digit number such as 12 requires two BCD symbols, just as in decimal:

12	=	0001	0010
		1	2

Since the microprocessor word is usually 1 byte (8 bits) we can just fit two BCD digits into each word. This is known as *packed BCD.*

Exercises

3-1. *Convert the decimal number 1022 to octal. The result should gladden the heart of patriotic thumbless Americans!*

3-2. *Convert the most significant byte of $A13F to decimal using first, the weight table method, and second, conversion via binary. Now, convert the entire hex number $A13F to decimal by either method. Question: Which do you find the easiest to use?*

3-3. *a) What range of signed integers (largest + to smallest −) can be represented in two's complement notation, using only one byte?*
b) What is the two's complement form of −128?

3-4. *a) What are the BCD codes of decimal 37, 42, 99?*
b) What is the decimal equivalent of BCD 0100 0111, 1010 0011? Are they both valid?

Unit 4: Binary Logic and Arithmetic

Unit 4

Binary Logic and Arithmetic

In this unit you will learn the rules of digital logic and how to apply them to perform binary arithmetic. These functions are used by the Arithmetic and Logic Unit (ALU) which is the operating heart of the microprocessor.

Learning Objectives—when you have completed this unit you should:

A. **Understand binary logic and truth tables.**

B. **Know how to apply this logic to implement a binary adder.**

C. **Know how the microprocessor performs addition and subtraction using two's complements.**

D. **Understand how the microprocessor employs logic function such as AND, OR, and X-OR.**

4-1. Binary Addition

The principle of binary addition is quite simple, in fact you already have used it in Sec. 3-1 (Complementary Notation), probably without even noticing it. But in order to understand the design of a binary adder in a microprocessor, let's state the rules explicitly.

Augend		Addend		Sum		Carry
0	+	0	=	0	+	0
0	+	1	=	1	+	0
1	+	0	=	1	+	0
1	+	1	=	0	+	1

Table 4-1. Binary Addition

Table 4-1 gives the rules for adding 2 bits in a single column of bits. The *carry* is the bit resulting from the 1+1 addition that should be "carried" over into the next higher order column of bits. The only unobvious instruction is what to do when a carry bit results from addition in the last and highest order column. Let's put this aside for the moment and consider only a single column.

A "black box" to add only one column is called a 1-bit adder or *half adder* and has the input-output functions shown in Fig. 4-1. Like any other black box, we need not know what is inside it but only that inputs result in the desired outputs as specified by the rules of Table 4-1. It is called a "half adder" because it does not do the whole job of binary addition. Table 4-1 tells you when to generate a carry for the next higher column but doesn't say what to do about an *incoming carry* from a lower-order column of bits. After learning how the half adder works, we will take up this question.

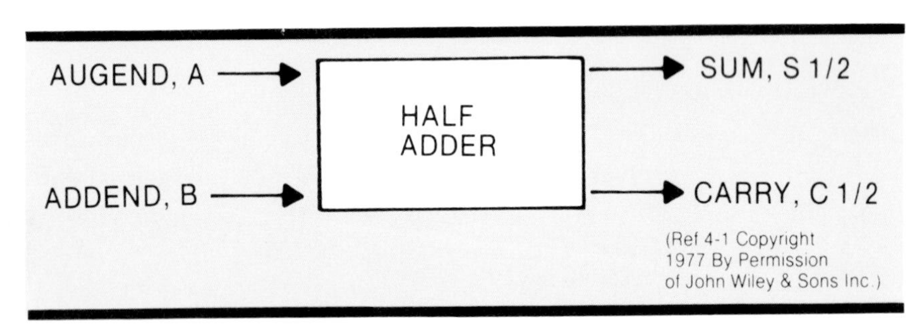

Fig. 4-1. Half Adder "Black Box"

The information of Table 4-1 can be put in a slightly different form called a *truth table*. A truth table specifies all the inputs and outputs of a logical box or system such as the half adder. The inputs, A and B, of Table 4-2 are the same augend and addend as shown in Fig. 4-1 and the outputs are the sum S 1/2 and carry C 1/2 of the half adder.

Inputs		**Outputs**	
A	B	S 1/2	C 1/2
0	0	0	0
1	0	1	0
0	1	1	0
1	1	0	1

Table 4-2. Half Adder Truth Table

Note that the regular patterns of the inputs A and B cover all possible combinations of 1 and 0 for the inputs. The outputs follow the specified rules of binary addition. Any combination of devices that results in the correct outputs as a function of the inputs will, by definition, be a half adder.

A truth table need not be restricted to two inputs or outputs. Any binary function can be specified by a truth table. If there are three input variables instead of only two there will be 2 to the third power or 8 possible combinations that must be specified; four inputs will require 2 to the fourth power or 16, and so on.

Exercise 4-1. *Write down in orderly fashion all the possible combinations of three binary input values. (Hint: look at the pattern scheme of A and B in Table 4-2.)*

4-2. Binary Logic

A binary logic device like a half adder is constructed of combinations of elementary logical functions, each of which can be considered as a black box with inputs and output(s) as shown in Fig. 4-2. Because there are four possible combinations of the two inputs of Fig. 4-2, there are 2 to the fourth power or 16 possible logic functions corresponding to 16 possible output columns. Some of these are trivial: for example, the output can always be 0 regardless of input or it can always be 1. (Perhaps these would correspond to a shorted switch or to an open circuit in an electrical "black box.") But the others are meaningful elementary functions that can be used to construct all other types of logic functions, including the adders and arithmetic units of microprocessors.

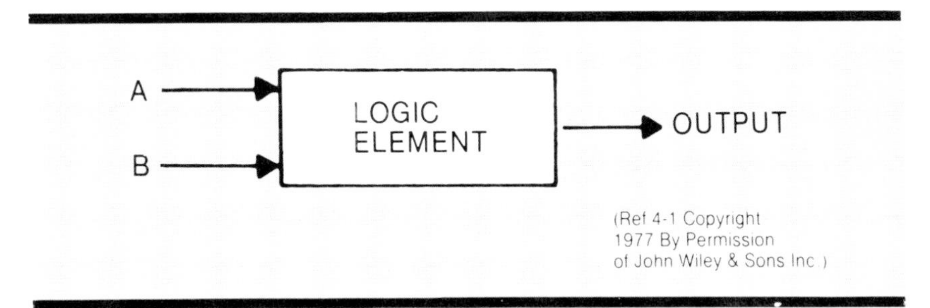

Fig. 4-2. General Logic Element

Exercise 4-2: *Construct a table showing all of the possible outputs columns corresponding to the elementary logical functions of two binary input variables. You should find 16 of these, including the "all 0" and "all 1" functions just mentioned. (Note also that the S 1/2 and C 1/2 outputs of Table 4-2 are two such outputs.)*

It can be shown that a small number, two or three, of these logical functions can be chosen as basic and can be used to build up any other possible elementary function, or even more complex logical devices, such as the half adder. The elements regarded as basic are somewhat a matter of choice, but one intuitively obvious set is that consisting of the AND and OR, together with the one-variable NOT function. Any possible logical result can be constructed from these three functions. Each of them can be described by a unique truth table and can be symbolized by a specific variation of Fig. 4-2.

4-3. AND, OR, NOT

The AND is symbolized by a specific "D" shape or by a multiplication sign (× or dot) as shown in Fig. 4-3. The truth table of AND is shown in Table 4-3.

Inputs		**Output**
A	B	C
0	0	0
1	0	0
0	1	0
1	1	1

4-3. AND Truth Table

Literally, the AND function is TRUE (meaning equal to 1) when *both* A and B are true (1). It is FALSE (0) when either or both inputs are zero (0).

One can imagine a hardware implementation of an AND function constructed by connecting two single-pole switches in series to a lamp and a battery, as shown in Fig. 4-4. If the two switches represent the A and B functions and are TRUE when closed, and if the lamp represents the output C and is TRUE when lit, then this circuit follows exactly the rules of the truth table, Table 4-3, and is therefore the representation of an AND.

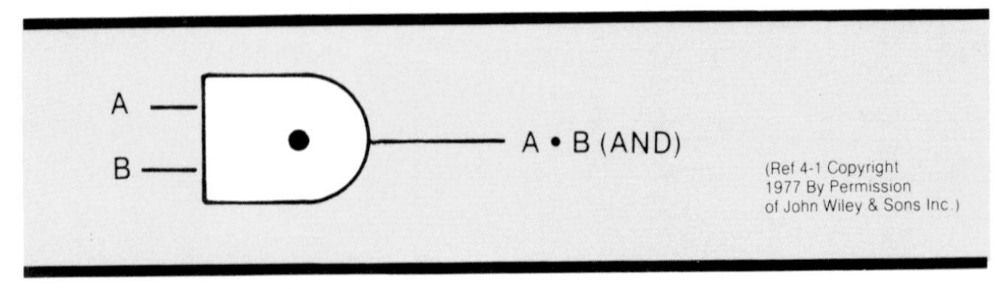

Fig. 4-3. AND Function Symbol

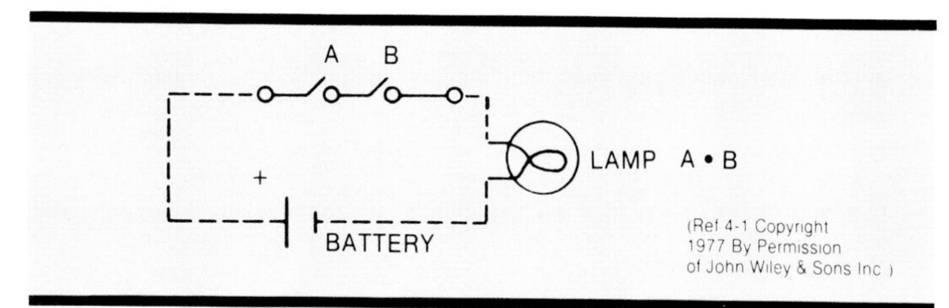

Fig. 4-4. AND Function Circuit Implementation

The OR is represented by a dished-in "D" and the symbol "+," as shown in Fig. 4-5. Its definition is the truth table, Table 4-4.

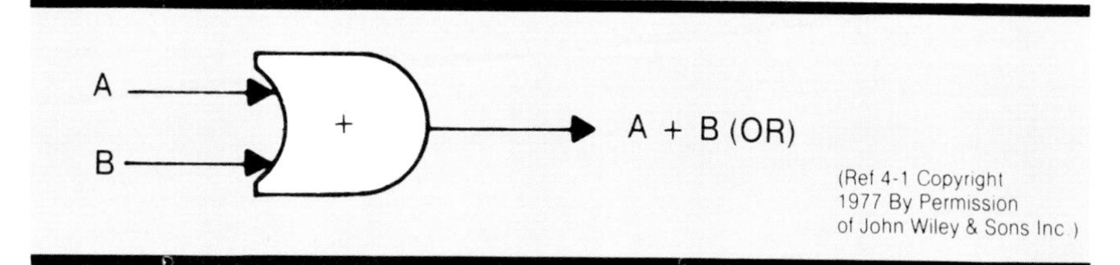

Fig. 4-5. OR Function Symbol

Inputs		**Output**
A	B	C
0	0	0
1	0	1
0	1	1
1	1	0

Table 4-4. OR Truth Table

As implied by its name, the OR is TRUE (equal to 1) when either the input A or the input B or both are TRUE (1). A circuit representing an OR can be made of two parallel switches, Fig. 4-6, that are substituted for the two series switches of Fig. 4-4. Closure of either switch or both will light the lamp.

Note that if we change the definition of TRUE to mean "lamp is not lit," that the OR circuit will behave like an AND while the AND becomes an OR. (Prove this to yourself by working out the truth tables.) This demonstrates that it is important to define exactly what is meant by TRUE, FALSE, 1 and 0 when analyzing logical circuits or devices. Practical integrated circuits (ICs) embodying logic elements such as AND and OR typically employ a high-voltage state such as 5 volts to represent 1 and a low voltage (less than 1 volt) to represent 0. This arrangement is termed *positive logic.* If the opposite convention is adopted, so that the TRUE state is low voltage or 0, it is called *negative logic.* Thus, we can change a positive logic OR to a negative logic AND using the same hardware, merely by adopting a different rule. Of course, the same rules must apply to all the logical elements that are connected together in any way.

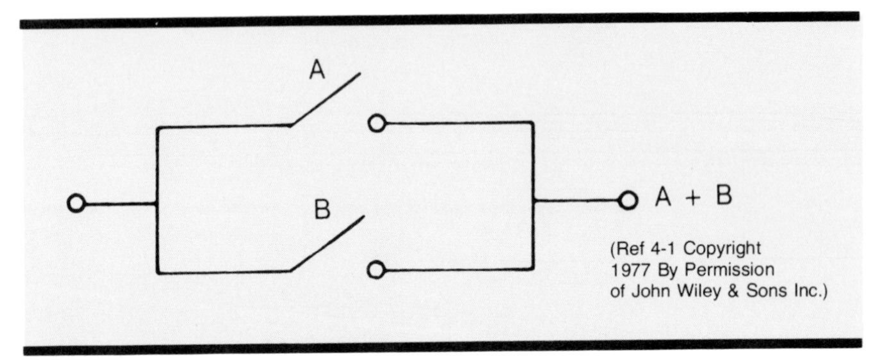

Fig. 4-6. OR Function Circuit Implementation

A special logic element is the NOT, since it is the simplest and has only one input and one output. The output is merely the inverse of the input, whatever that may be, 0 or 1. Thus, the NOT has the truth table shown in Table 4-5.

Input	Output
0	1
1	0

Table 4-5. NOT Truth Table

The NOT function is symbolized by a small circle or by a triangle representing a one-stage (inverting) amplifier as shown in Fig. 4-7.

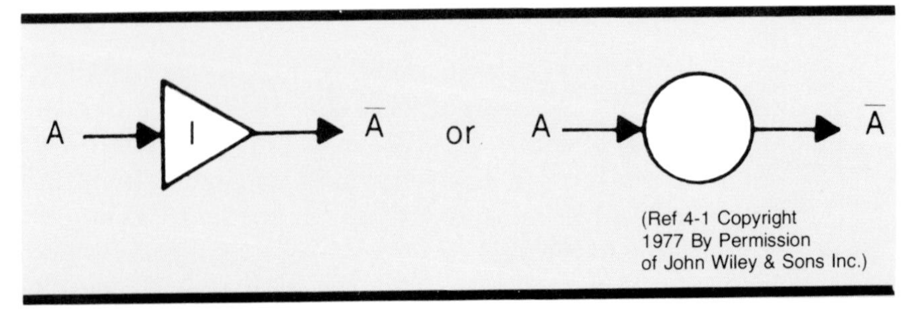

Fig. 4-7. NOT Function Symbol

4-4. NAND, NOR Functions

All of the 16 possible functions of two inputs you find in Exercise 4-2 can be assembled from some combination of AND, NOT, and OR elements. For example, an AND followed by a NOT (see Fig. 4-8) can be defined by the following truth table.

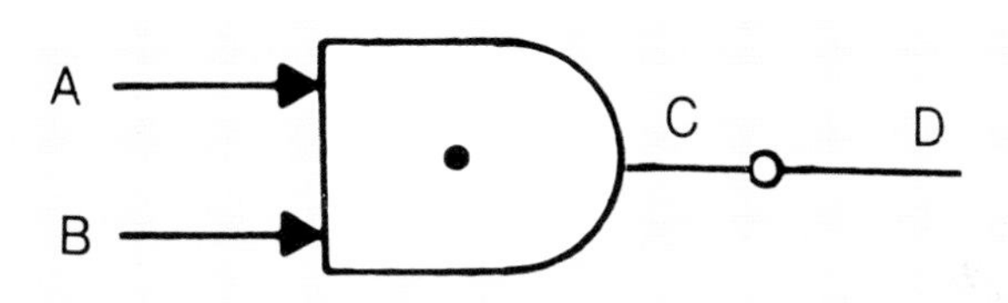

Fig. 4-8. NOT-AND Logic (See Example 4-1)

Example 4-1
See Fig. 4-8.

Inputs		AND OUT	NOT-AND
A	B	C	D
0	0	0	1
1	0	0	1
0	1	0	1
1	1	1	0

The result is called "NOT-AND," abbreviated "NAND." If the first element had been an OR instead of AND we would have constructed a "NOT-OR" or "NOR" device. (See Fig. 4-9 and Fig. 4-10 for symbols of these devices.) The NAND and NOR themselves can be considered basic elements; just as the NOT, OR, and AND can be used to construct all other binary logic properties, so can the two elements NAND and NOR. Multiple NAND and NOR elements are very commonly implemented in MSI (medium-scale integration) IC's because of this versatility. However, you should note that it usually requires more NAND and NOR elements to construct some logical function if we restrict ourselves to these two, than if we permit the use of AND, OR, and NOT elements as well.

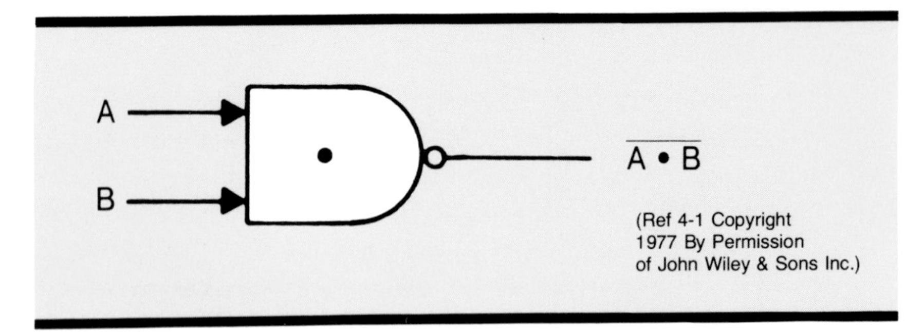

Fig. 4-9. NAND Symbol

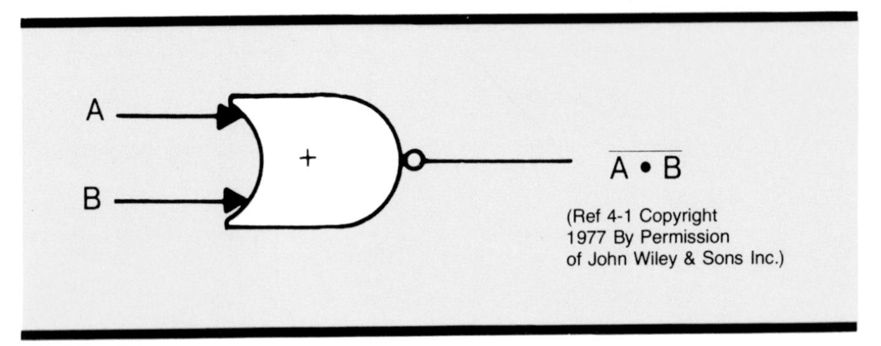

Fig. 4-10. NOR Symbol

Exercise 4-3. *Construct a NOR truth table.*

Exercise 4-4: *Three elements, two ANDs and an OR, plus two NOTs, are connected as in Fig. 4-11. Develop a truth table for this function. (The result will be an important function used in the half adder, as we shall see in the next section.)*

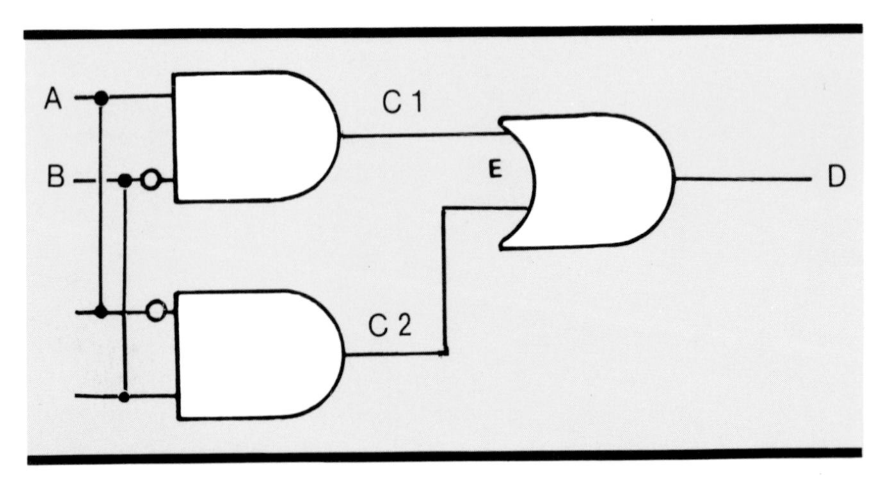

Fig. 4-11. See Exercise 4-4

4-5. Half Adder and X-OR

We have now put together the background to analyze and implement the half adder, Table 4-2, as we intended. Actually, there are many different logic combinations that will result in this function. Figure 4-12 is one that is economically composed of four elements, two ANDs, an OR, and a NOT.

To prove that Fig. 4-12 is indeed a half adder all we need do is construct its truth table and compare the outputs with Table 4-2. These details are worked out in Table 4-6, which you should study carefully to be sure you understand what is happening at each step.

A	B	D (OR)	E (NOT AND)	S 1/2	C 1/2
0	0	0	1	0	0
1	0	1	1	1	0
0	1	1	1	1	0
1	1	1	0	0	1

Table 4-6. Proof of Half Adder Logic (Fig. 4-12)

The output D is merely the OR of the A,B inputs. E is the inverse (NOT) of the AND with A,B inputs. E could have been produced by a NAND, but we also need the AND output for a carry, C 1/2. The sum output, S 1/2 is the AND of D and E. It is identical to the S 1/2 column of Table 4-2. Likewise, the output of the first (A,B) AND is identical to the carry C 1/2 of Table 4-2. Therefore, the device is a half adder.

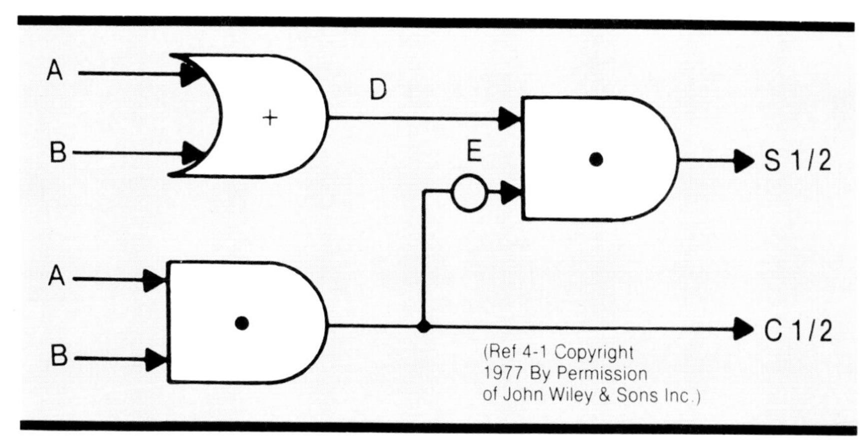

Fig. 4-12. Half Adder Logic Implementation

The S 1/2 output is one of those you will obtain in Exercise 4-1; it is itself an elementary logical function. It is known as the Exclusive-Or or X-OR. the symbols for the X-OR are shown in Fig. 4-13. Its truth table can be copied from the A,B, and S 1/2 columns of Table 4-6. Note that it is identical to the result of Exercise 4-4, Fig. 4-11. Therefore, an X-OR is equal to the OR of two AND functions, as in Fig. 4-11.

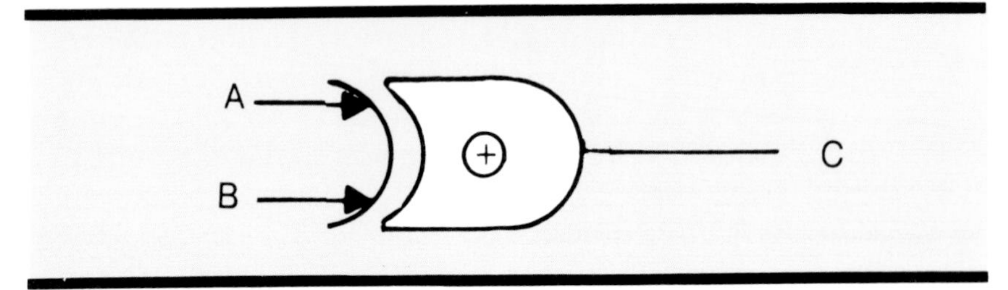

Fig. 4-13. Exclusive-OR (X-OR) Symbol

4-6. Boolean Algebra

An alternative method of describing logical circuits by means of algebraic equations or expressions exists. It is called *Boolean algebra* after an English mathematician and logician, George Boole (1815-1864), who first explained it in his book *The Laws of Thought*. A similar method is called the *propositional calculus*. These algebraic expressions often are used in catalogs and the literature of logic ICs, so they will be explained briefly here.

Various symbols are used, but one common set is the + for OR, the multiplication sign (×, dot, or merely adjacent symbols such as AB) for the AND function, and an overbar or the swing dash (~) for the negation or NOT. The X-OR is symbolized by a circled plus sign (⊕).

The algebraic method of proving validity of a logical formula often is quicker and easier than constructing exhaustive truth tables. If a formula can be reduced to an "obvious truth," called a *tautology*, then the logical formula is true in its original form. Boolean algebra can be and is most often used to reduce the complexity of logical expressions and so simplify the hardware necessary to implement them in an electronic circuit.

Some tautologies (some really obvious, some not so obvious, but all of which can be proven by truth tables) are:

Double negation $p = \sim(\sim p)$
Excluded middle $(p + \sim p) = 1$ (always true)
Contradiction $p(\sim p) = 0$ (always false)
DeMorgan's theorem $\sim(pq) = \sim p + \sim q$

In addition to the above, the normal laws of algebraic manipulation with regard to parenthesis, commutation, and association apply to Boolèan algebra.

For instance, using algebraic expressions, we can describe the X-OR arrangement of Fig. 4-11 as follows:

Example 4-2:

Algebraic proof of X-OR
A AND NOT B; $C1 = A(\sim B)$
B AND NOT A; $C2 = \sim A(B)$
C1 OR C2; $A(\sim B) + \sim A(B) =$ X-OR

Example 4-3

Algebraic proof of half adder, see Fig. 4-12.
$S\ 1/2 = DE$; $D = A + B$; $E = \sim(AB)$
$E = \sim A + \sim B$ (by DeMorgan's theorem)
$S\ 1/2 = (A + B)(\sim A + \sim B)$ (multiply out terms)
$= A(\sim A) + A(\sim B) + B(\sim A) + B(\sim B)$
$= 0 + A(\sim B) + B(\sim A) + 0$ (by contradiction)
$S\ 1/2 = A(\sim B) + B(\sim A) =$ X-OR
and $C\ 1/2 = AB$ (by inspection)

4-7. Full Adder

The addition problem for a column of bits is only partially solved by the half adder. The full adder not only must generate a carry bit for the next higher column but also must know what to do with a carry bit coming in from the lower column. Thus, the full column adder has three inputs—the augend and addend as in the half adder and the carry bit from the next lower stage. The outputs, as before, are the sum S and the carry C1. If at least two of the inputs are 1s, a carry is generated. Thus, the full adder can be cascaded for any number of steps to provide simultaneous addition of multiple columns of bits or *words*, such as a byte. The action of the full half adder is summarized in Table 4-7.

Inputs			Outputs	
A	B	C	S	C1
0	0	0	0	0
1	0	0	1	0
0	1	0	1	0
1	1	0	0	1
0	0	1	1	0
1	0	1	0	1
0	1	1	0	1
1	1	1	1	1

Table 4-7. Full Adder Truth Table

Exercise 4-7 points up a limitation of binary machine addition as compared to the usual pencil and paper method. If a carry is generated in the highest order of bits it will *overflow* the adder register. There will be no place to put the overflow carry bit so the answer will be erroneous. In a practical machine, it is necessary to recognize the overflow condition and to take the necessary steps to handle it and prevent error.

The statements made at the start of this section would permit one to write logical equations for S and C1 like those in Example 4-3.

Exercise 4-5: *Using the full adder table, add in binary the equivalent of decimal 29 to decimal 25.*

Exercise 4-6: *Assuming an 8-bit adder, sum the binary equivalents of decimal 45 and 145.*

Exercise 4-7: *Do the same for the binary equivalents of decimal 255 (11111111) and decimal 2. Remember you have only an 8-bit adder. What is the answer? Is it correct? (reconvert to decimal) Why or why not?*

An adder design could be generated from these equations. We'll not take the space to do this here, but instead note that the result shows, not surprisingly, that a full adder can be obtained by putting together two half adders. Again, there are many ways to implement this result in hardware. Figure 4-14 shows only one method.

Exercise 4-8. *Write the truth table for the full ladder logic (neglecting the dashed lines and box) and assume that the lower order carry C comes directly into the second half adder. Have you proven it is an adder? How?*

The dashed box C in Fig. 4-14 shows how it is possible to use a single adder circuit in time serial fashion to add multiple columns of bits. The inputs (A,B) for each column, beginning with the lowest order, are presented to the adder in sequence. The outgoing carry for each column is saved in a temporary 1-bit memory (the dashed box), and is presented in an incoming carry when the next higher order A,B pair is entered. This *ripple method* is seldom used in contemporary microprocessors because of the additional processing time to perform the entire multibit add. However, as we will show in a subsequent example, it is similar to the principal of multibyte precision arithmetic.

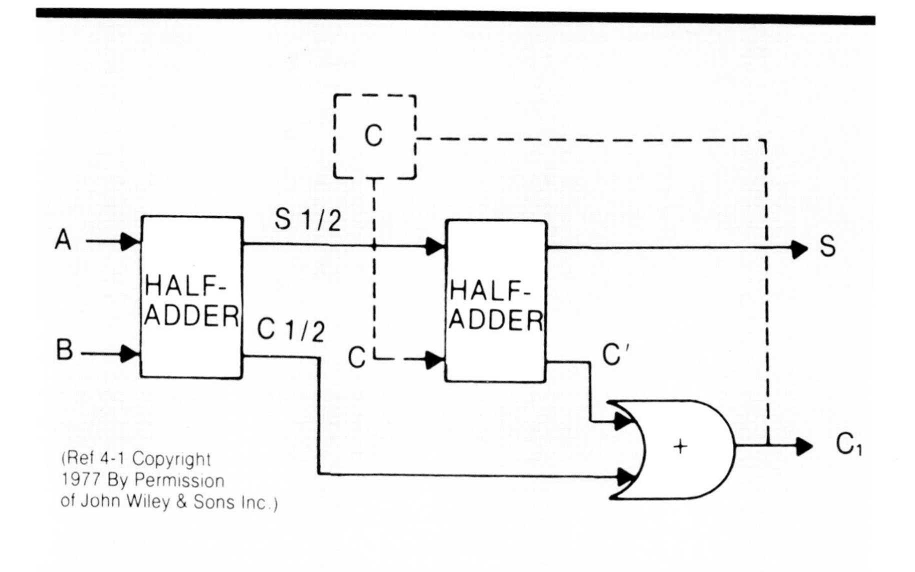

Fig. 4-14. Full Adder Logic Implementation

4-8. Two's Complement Addition and Subtraction

In the last unit (Sec. 3-7) we stated, without proof, that the two's complement form of a number could be employed for subtraction, using normal adder logic and hardware. You will recall that using two's complements automatically takes care of the sign and magnitude problems. In this section we will demonstrate how this is accomplished.

Inspection of the table of two's complements for 4-bit numbers (Table 3-3) will remind you that the highest order bit reveals the sign of the number. If the high order bit is 0, the number is positive, and if it is 1, the number is negative. In most commonly used microprocessors (such as the 6502, 8080, 6800, and Z80), the registers performing arithmetic will handle only 8 bits or 1 byte at a time. Thus, for a *signed* number we only have 7 bits available. The largest positive number that can be held in an 8-bit register is then +127 (0111 1111), and the most negative number is −128 (1000 0000). (At this time, we only are considering integers. Thus, the radix point is fixed and is assumed to be to the right of the rightmost bit (LSB).

When adding two positive (signed) numbers in two's complement there is no difficulty when the sum does not exceed 127.

Example 4-4
Add 65 and 60 (decimal).

```
0100 0001        65
0011 1100        60
---------        ---
0111 1101        125 (+)
+
```

The sign of the result is + because the left most bit is 0.

Note that when the sum is greater than +127 or −128, overflow occurs. This is because the 8-bit register is not large enough to hold the true sum. This condition is always signified by the sign bit changing.

Example 4-5
Add +65 and +65 (decimal).

```
0100 0001        65
0100 0001        65
---------        --
1000 0010        ?
OVERFLOW! (Sign bit changed from 0 to 1)
```

In most microprocessors, an *overflow flag* is set when this condition occurs. The "flag" is a bit in a special register called the *status register*.

Subtraction involves changing the sign of one numer (called subtrahend) and adding it to another number (minuhend). The change in sign of the subtrahend is accomplished by taking the two's complement.

Example 4-6
Subtract 65 from 127 (decimal).

```
  0100 0001      +65
  1011 1110      one's complement
         +1      add 1
  ---------
  1011 1111      two's complement
  0111 1111      +127
  ---------
1 0011 1110      SUM = 62 (decimal)
* Extra carry
```

Note that an extra carry is generated which is ignored and discarded. The sign bit (bit 7) remains 0 since the result is positive. No overflow is possible with the addition of a positive fixed point number to a negative one.

Example 4-7
Subtract 67 from 65 (decimal).

```
0100 0011      +67
1011 1100      one's complement
       +1
---------
1011 1101      −67
0100 0001      +65
---------
1111 1110      SUM = −2
```

The sign bit of the sum is 1, so the result is negative. That it is equal to −2 (decimal) can be seen by comparing with Table 3-3 in Unit 3, or by taking its two's complement.

Exercise 4-9. *Take the two's complement of the result of Example 4-7 to obtain its (unsigned) value.*

When two negative numbers are added, we sum their two's complements.

Example 4-8
Subtract 16 from −6 (decimal).

```
      0001 0000        +16
      1110 1111
              +1
      ---------
      1111 0000        −16
      1111 1010        − 6
      ---------
    1 1110 1010        SUM = −22
    * (Discard extra carry)
      0001 0101
              +1
      ---------
      0001 0110        two's complement of sum = +22
```

Overflow is possible in this case when the magnitude of the sum of the negative numbers exceeds (−)128.

Exercise 4-10. *Subtract 65 from −64 (decimal). What is the result? Why?*

In the example above we discarded the extra carry, if any, out of bit 7. Bits are often numbered from the rightmost, starting with 0, so that bit 7 is the most significant bit, MSB, or sign bit in these examples. Bit 0 is then the least significant bit (LSB). In many microprocessors the carry out of bit 7 is preserved in the status register. It can be used to perform 16-bit (double precision) arithmetic by carrying it over as an input to the register holding the higher order byte sum. Just as we showed earlier that a single bit adder can be used sequentially to add multiple columns of bits by saving the carry for each column and recycling it, Fig. 4-13, so it is possible to sequence an entire 8-bit register, using it first for the lower-order byte and secondly for the higher-order byte or bytes.

4-9. AND, OR, X-OR Bit Operations

Addition and subtraction are two of the fundamental operations of the microprocessor, and, combined with *shift* of the entire register contents (byte) right or left relative to the binary point, can accomplish multiplication and division as well. (This will be taken up in a subsequent unit.) But many of the logical functions are of themselves important in microprocessor operations. We already have seen how the NOT is used for one's and two's complement. The AND function is also useful in many ways. For example, if we wish to test whether or not a specific bit is a 1 or a 0, we need only AND it with a byte having a bit 1 in that position.

Example 4-9
Given the byte 1010 101? where ? is either 1 or 0, how can we find the value of the unknown bit?

Answer. AND unknown with "*mask*" 0000 0001 ($01) and test bit no. 0.

If ?=1

```
1010 1011
0000 0001      AND with $01
---------
0000 0001      Result, bit 0=1
```

The result is 1 in test bit number 0 so that ? must be 1.

If ?=0

```
1010 1010
0000 0001      AND with $01
---------
0000 0000      Result, bit 0=0
```

Since the result in LSB (bit 0) is 0, the value of ? must be 0.

(If this is not perfectly clear, review Table 4-3 and note that if column A is the unknown bit and B is 1, then the output of the AND circuit is 0 when and only when A=0.)

Another obvious use of AND is to *clear* a register (set all bits to by ANDing it with $00 (binary 0000 0000).

The OR can be used to *set* bits, that is, to insure that they are equal to 1. Assume that we want to set the four lower bits of a byte to 1111, then we OR it with $0F, as in the example below.

Example 4-10
Convert 1010 1010 to 0000 1111.

1010 1010	Original byte
0000 1111	AND with mask $0F for upper nybble
0000 1010	Result (clear upper nybble)
0000 1111	OR with same mask for lower nybble
0000 1111	Result (set lower nybble)

The Exclusive-OR (X-OR) is less frequently employed for bit operations, but it can be used for one's complementing by X-ORing all bits in a register with $FF.

Example 4-11
One's complement the byte $4A.

0100 1010	$4A
1111 1111	X-OR with mask $FF
1011 0101	Result (one's complement)

The X-OR can also detect the change in status of a bit since an X-OR with 1 changes from 0 to 1 when the bit itself changes from 1 to 0.

References

[1]Bibbero, R.J., *Microprocessors in Instruments and Control*, New York: John Wiley (Wiley-Interscience), 1977, Chapter 3.

Unit 5: Memory, Addressing, and Data Buses

Unit 5

Memory, Addressing, and Data Buses

In the previous unit you learned some useful operations on data, logic, and arithmetic that a microprocessor can be instructed to perform. Ultimately, these operations are combined to accomplish industrial and process control tasks. For the most part, they are done by the Arithmetic and Logic Unit (ALU) of the microprocessor. But the microprocessor unit (MPU hereafter) consists of more than just the ALU. There are other units, such as memory (devices that store data), that must be added to make the MPU into a microcomputer or controller. Data and instruction words (information collectively) must be moved in and out of the ALU from these other devices and from other parts of the MPU so that the useful operations can be performed.

Learning Objectives — When you have completed this unit you should:

A. **Know the structure of microprocessor memory devices and how they are organized.**

B. **Understand the concept of** *address space* **and how information is moved and stored.**

C. **Know why address and data buses are used and how they are employed by the MPU.**

D. **Comprehend the function of decoders and control signals in the movement and storage of information.**

5-1. Storage of Information

A *microprocessor*, or MPU, is a small, inexpensive, mass-produced machine used to process binary data at extremely high speed. The low cost, a few dollars, is achieved by the technique of printing microscopic transistor circuits on very small pieces of specially treated silicon coated with metal (aluminum) and a nonconductor (silicon dioxide). The silicon after treatment is called a *chip* and the process is termed large-scale integration (LSI) because tens of thousands of transistor circuits can be printed on each chip. The active circuits, known as *gates* can store a binary 1 or 0 and can, when required, alter the stored value. A schematic cross section of a chip and its associated gates is shown in Fig. 5-1.

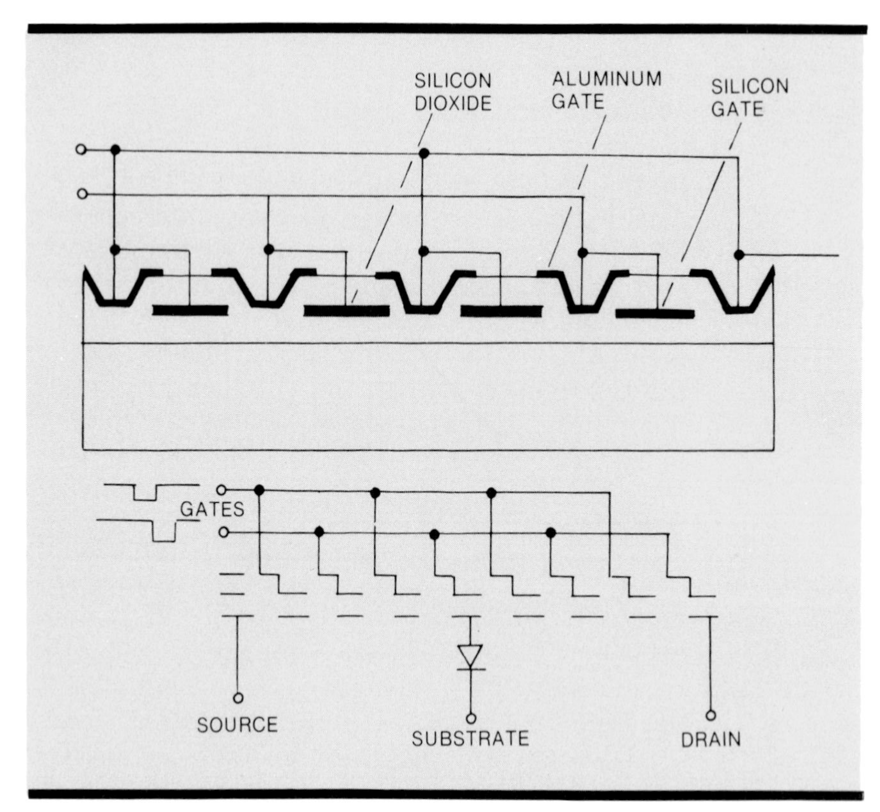

Fig. 5-1. Cross Section of LSI/MOS Chip Showing Gates

The specific means of producing LSI chips varies, but for the most part, microprocessors and the "families" of chips associated with them are formed by processes known collectively by the name of MOS for *metal oxide silicon.* There are different kinds of MOS technologies. For example, PMOS uses silicon *doped* (or impregnated) with impurities that produce positive charges or "holes." This process is relatively old and inexpensive, but also causes slower operation of the circuitry. NMOS chips are doped to produce excess electrons to carry the electric charges more rapidly than holes, but the process is more expensive than PMOS. CMOS (complementary MOS) combines both types of charge carriers, and though slower than NMOS, needs very little power to operate. *Bipolar* is another technology, very much faster than MOS, but also expensive and power-consuming. It is generally used on large main-frame computers where cost and power are not significant considerations.

MOS chips can be fabricated as microprocessors or nearly any other solid-state electronic device. The simplest and earliest LSI devices to be mass produced in MOS were *memory chips*, used for storage of data and program information. The first memory chips in the new technologies were fabricated into a small number of 1-bit cells, generally 256, and were gradually made larger until now 16K-bit (16,384) and 64K memories are common and 256K units are coming into use. (Note that memory-chip size is measured in bits, not bytes or words.)

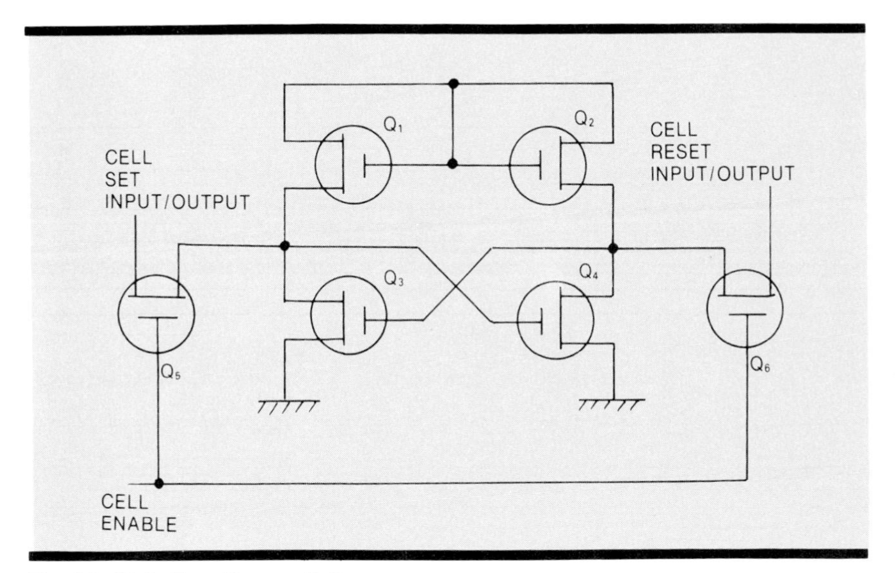

Fig. 5-2. Static Ram Memory Cell

Figure 5-2 shows the schematic of a single memory cell, holding 1 bit and constructed of six transistors. The four inner transistors are connected as a *bistable* or *flip-flop* circuit, one which has two stable states representing "on" or "off," 1 or 0. The cell outputs a 1 when it is in the set state and a 0 when reset. The flip-flop will hold its state as long as the cell has power and so is called a *static RAM*: a random access memory that is stable under static or direct current (dc) conditions. When a set or reset signal is allowed to enter, a 1 or 0 is *written* into the cell, consequently it is also known as a *read/write* or R/W memory.

A simpler memory cell can be made of a single transistor gate plus a storage capacitor that takes advantage of the high impedance of MOS and its ability to prevent the capacitor charge from leaking off too rapidly, see Fig. 5-3. The presence of the charge then represents a 1 in the cell. But eventually this charge will leak off and must be replaced, even though power to the memory chip is continuous. The 1 value of the bit must be continually recycled into the cell, hence it is called a *dynamic Ram* or DRAM; contrasted with the static type. The DRAM may be written to as well as read, so it is also a read/write memory.

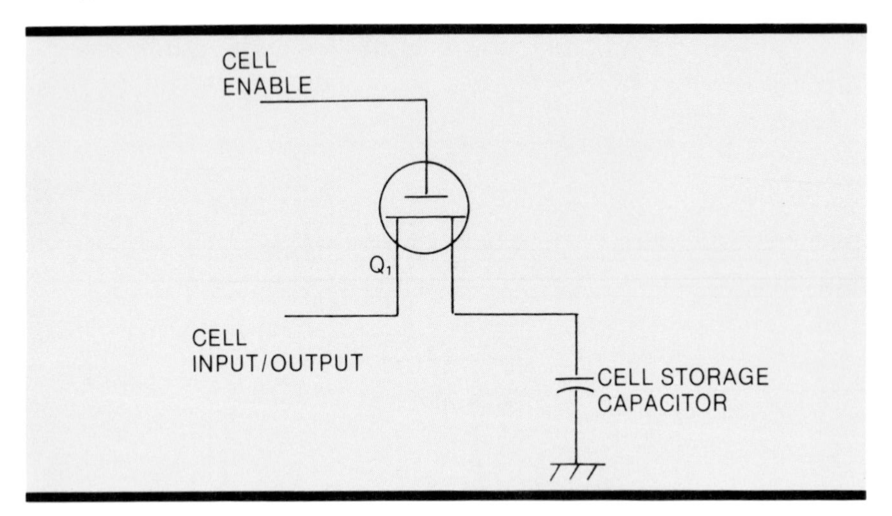

Fig. 5-3. Dynamic Ram Cell

There are many other kinds of memories used in microprocessor systems. The RAM is called random access because any cell in the memory can be read out or written to in approximately the same length of time. Another kind of random-access memory is the ROM or *read-only memory*. The ROM can only be read, not written to. Consequently, it is desirable to use ROM for parts of memory that never change, such as an operating program for a PID controller, while RAM is used for rapidly changing data or intermediate calculation results. ROM units are less expensive to manufacture in large quantity but there is a fixed cost for any specific bit pattern or program that is built into the memory, as the result of the customized processing step known as *masking*. This last step in manufacture actually determines the pattern for a specific customer, so it is impossible to stock completed ROMs or take full advantage of the economics of scale.

Another type of read-only memory, called the *field-programmable ROM* or PROM can be programmed by the user without a mask. A higher-than-normal voltage is used to "blow" fusible links connecting gates so that logic ones become logic zeros in the pattern desired by the user. PROMs, once programmed, cannot be changed again, but there is yet another type known as the *erasable PROM* or EPROM; the pattern programmed into these can be erased by ultraviolet light, "healing" the blown fuse links, and resetting the EPROM so that it can be reprogrammed. Thus the EPROM is very convenient to use during development of a system when it is probable that the program will have some errors and require correction. After the system is well "debugged," the EPROM can be exchanged for PROM or ROM, whichever is justified by the quantities needed.

A newer type of reprogrammable ROM should be mentioned as it may find greater use in the future. This is the *electrical-erasable PROM* (EEPROM), also called the *read-mostly memory* or RMM. Like the RAM, it is possible to write as well as read to the RMM, but with these differences: the write speed is much slower than with RAM, but the bit pattern is retained indefinitely without power, as with ROM or PROM. Thus, the RMM is useful for storing information that must be changed occasionally by an operator, such as the configuration of a process controller, but will remain constant during normal use of the machine.

The major difference between the RAM and the various kinds of ROMs is that the information bits in the former are *volatile* and will "evaporate" when power is removed. Therefore all programs that must be used again and again, every time the equipment is put into service, must be in nonvolatile storage such as ROM or the magnetic types discussed next.

Magnetic memories are used with some microprocessor systems. This technology is much older than the semiconductor types we have been discussing. Early computers used *magnetic core* memory almost exclusively for main memory. The magnetic core is a miniature loop of ferrite, a ceramic with strong magnetic powers, that is so threaded with fine wires as to be capable of magnetization in either of two directions. The direction of magnetism is used to signify the presence of a logic one or a zero in the core. A core memory system consists of perhaps 4K cores wired in such a way that they can be randomly accessed (read or written to). The magnetic core memory is nonvolatile because the direction of magnetism is retained when power is removed from the system. The magnetic core memory is expensive in terms of cost and power and has been almost entirely supplanted by MOS technologies, especially in microcomputers. There are, however, still some uses for it, for example, the retention of key status information in the event of a sudden power outage. (The Honeywell TDC 2000 Basic Controller uses core memory in this way to store configuration information for backup or restart.)

The *magnetic tape* and *magnetic disk* are much more commonly found with microcomputer systems. Magnetic tape differs strongly from the types we have just discussed in that it is a *serial access* rather than a random-access storage medium. The bits on a tape are stretched out in a line and must be read in sequence, thus the information at the end of the tape is accessed much later than the information at the start. An example is an audio tape cassette, which is usually played from one end to the other in a fixed direction. In fact, audio tape cassettes are commonly utilized for digital program storage with personal computers. (The bits are recorded in the form of tones, the frequency change indicating the difference between 1 and 0.) Magnetic tape is a slow means of accessing particular blocks of information because of this serial feature (and because of relatively slow recording and reading speed in the case of converted audio cassette players). Therefore, it is used largely for the storage of permanent files of data or programs that are loaded into main memory at the start of a session.

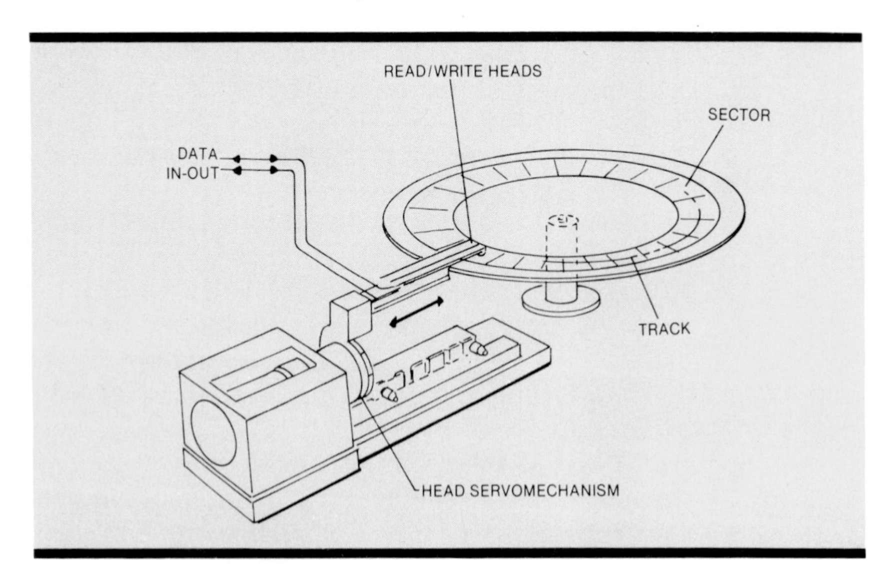

Fig. 5-4. Magnetic Disk Memory

Magnetic disk memories are a considerable improvement both in access speed and convenience. The chief form of disk memory employed with microcomputers is the flexible removable disk, called a *floppy disk*. This disk is made of plastic, coated with magnetic material, and is spun under a set of *read and write heads* that are capable of magnetizing spots on the disk in either a 1 or 0 direction and reading back the data. As shown in Fig. 5-4, the information is stored in concentric circles called *tracks*, each track being further segmented into arcs called *sectors*. The head is precisely positioned over its tracks by an electrical servomechanism or stepping motor, and the disk rotates under the head until the desired sector appears. The sector bits are then read serially. Since the tracks are accessed (almost) randomly, and the sectors are read serially, the magnetic disk is somewhat of a compromise between random and serial access. The type of disk called *minifloppy* is 5.25 inches in diameter and can hold anywhere from about 80K to 2 megabits. It can find a given sector in a few hundred milliseconds and transfer information at the rate of 125K bits/seconds. Larger floppy disks (eight inches) are more common in commercial machines, such as word processors, and are faster and hold proportionally more information. The most costly and effective type of disk system used with microcomputers is called the *Winchester disk*, presumably the original code name for the International Business Machines 3340 disk system (now manufactured by many others). The Winchester disk is rotated in a sealed environment and has a lower head height and mass, which permit much greater writing density and rotational speed. The result is much more storage capacity (about 8 megabytes per side for an 8-inch disk, over 6 megabytes for a 5.25 inch) and access times averaging 25 to 70 milliseconds (170 ms for the 5.25 disk).

5-2. Movement of Information

The MPU is only an information processor. It does not have any useful function unless it connects with something in the outside world, such as an industrial process, a machine, or a person. So it must be able to accept new data and put out information in the form of control signals, displays, or reports. Figure 5-5 illustrates the need to connect the MPU with an input (a person or a sensor), an output (a person or a process), and a memory store. The memory is necessary because the *program*, that instructs the MPU what task to perform and in what order, and the data, that may represent initial conditions, intermediate calculation, or final results, have to be held somewhere, and there is not room on the MPU chip. (If the MPU should contain some RAM or ROM it is then called a *one-chip microcomputer* rather than a microprocessor, but the current practice is to employ separate chips as the *main memory*.) The program of instructions can be stored in ROM, RAM, or PROM and be equally accessible to the MPU since the addressing schemes are the same in all three. Magnetic media, such as tape or disk, are *not* part of the main memory (for microprocessors) because their slower access time would hold back the MPU and force it to wait for memory transfers. They form a separate class called *mass memory*. Access from mass memory to main memory is usually controlled by the MPU through input and output ports, so that the mass memory can load a program or data into the RAM and a revised program or new data can be stored or *saved* permanently in the mass memory. (In some microprocessors, as we will see in the next unit, it is possible to transfer data directly from mass memory and other external devices into the RAM by using the data and address buses without the MPU, using a technique called *direct memory access* or DMA.

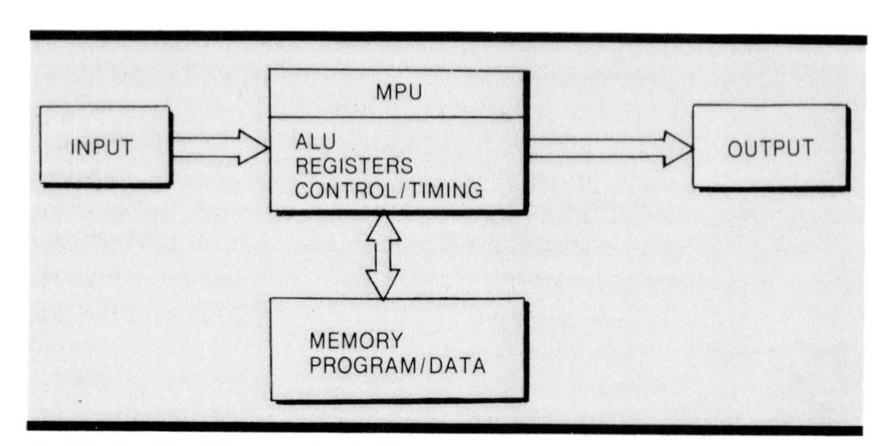

Fig. 5-5. Simplified Microcomputer Architecture

The MPU, Fig. 5-6, consists of the ALU, the *accumulator* that holds its output, a counter, temporary word-storage devices called *registers*, and units for timing and control. The timing unit keeps all parts of the MPU working in synchronism under the control of a quartz crystal *clock* external to the chip. The control unit interprets the instructions in the program and decodes them into simple commands (called *microcode* for each part of the ALU, registers, and counter. The register storage units are capable of holding 8 or 16 bits (in an 8-bit microprocessor). Thus, they may hold one byte of data or part of an *address*. (Registers may be available to the programmer, in which case they are known as *general registers*, or they may be dedicated to special purposes.)

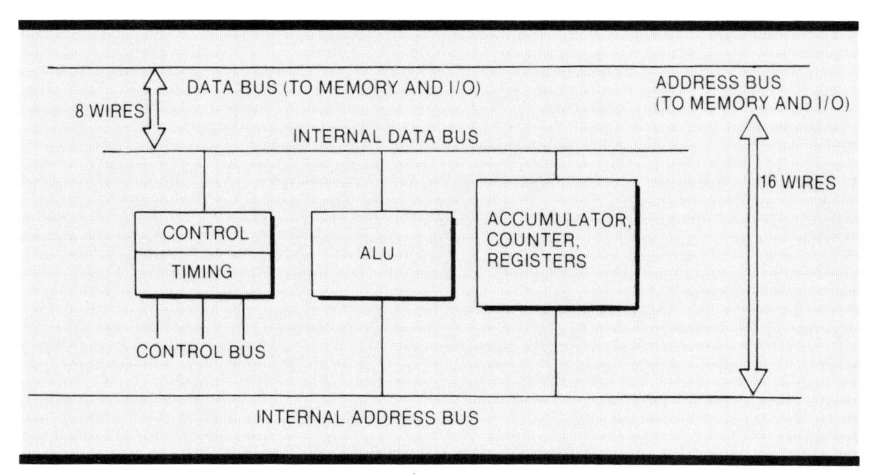

Fig. 5-6. Microprocessor Unit Bus Architecture

The address is a code stating where in memory (or from which ports) data or program instructions are to be found and brought into the MPU (*fetched*), or to which memory cells or ports it is to be written. The *address concept*, together with the data and address buses are two of the most important ideas in computers. The address bus enables data and instructions to be moved rapidly in and out of a very large random-access memory. In turn, this makes practical the idea of a *stored program* of instructions that can be retrieved rapidly enough to keep pace with the sequence of microprocessor actions. The sequential stored program is the core of the original idea of the modern computer, known as the *von Neumann machine* after its inventor, mathematician John von Neumann. The rapid-access random memory also makes practical the concept of *program branching*. This is the ability to introduce an element of *decision* into a computer program that otherwise would consist merely of a fixed sequence of actions. Branching allows the program to *jump* out of sequence, to a new program segment or back to repeat an old one, purely on the basis of new data or calculation results. In plain words, the computer can be instructed to behave as follows: "If x is greater than y, take the action described after this segment, otherwise, continue with the remainder of the program."

Figure 5-6 illustrates how bytes of data are moved between the MPU devices through the channels we have called buses. On the chip itself the buses consist of a number of conductors, one for each bit in the word to be moved. Collectively they are called an *internal bus*. When the information must leave the confines of the MPU chip and move to memory or other devices, the buses consist of wires or printed circuit conductors, and are termed the *data* or *address buses*, depending on their use. The third collection of conductors shown on Fig. 5-6 is the *control bus*; it carries synchronizing and control signals that will be discussed in a subsequent unit.

The data bus moves, simultaneously and in parallel, all the bits that make up a piece of data, for example, from the ALU to a register or from register to memory. Since an 8-bit MPU deals with 1 byte at a time, there must be eight conductors to the bus. It is important that all eight move and arrive at their destinations at the same time (within one clock cycle), otherwise the word may be garbled.

5-3. Addressing and Address Space

To store and process large programs, an 8-bit microcomputer should be able to address 64K bytes of random-access R/W storage directly, or a total of 524,288 bits. Obviously, we are not going to connect all of these bit cells to the MPU with a half-million wires! In fact, it is very important to keep the number of connections into the MPU at a minimum, for two reasons. First, every external connection is a potential source of failure and so decreases reliability. Second, the number of pins on a chip connecting to the circuit board is severely limited for economic reasons, usually to 40 or less. All data, addressing, control signals, and power must go through these 40 pins.

To see how we can whittle down the number of connections, let's take another look at Fig. 5-2 and Fig. 5-3. Each memory cell (bit) has an input/output terminal which permits the cell contents to be read or written to. The output voltage is high if the cell contains a 1 and is low if its state is 0. In addition, each cell has an enable line which must be specifically activated if writing is to take place.

Now consider the arrangement of many such cells in a memory chip. Figure 5-7 shows a hypothetical arrangement of 1024 (1K) cells arranged in 32 rows and columns. In order to read out or write into any cell, an electrical connection to its I/O terminal is needed. But instead of a contact to each individual cell, let's assume we can make a partial connection to an entire row at one time. To complete the I/O connection, it is only necessary to activate the particular column in that row. By the same token, all cells in a column have the second connection completed

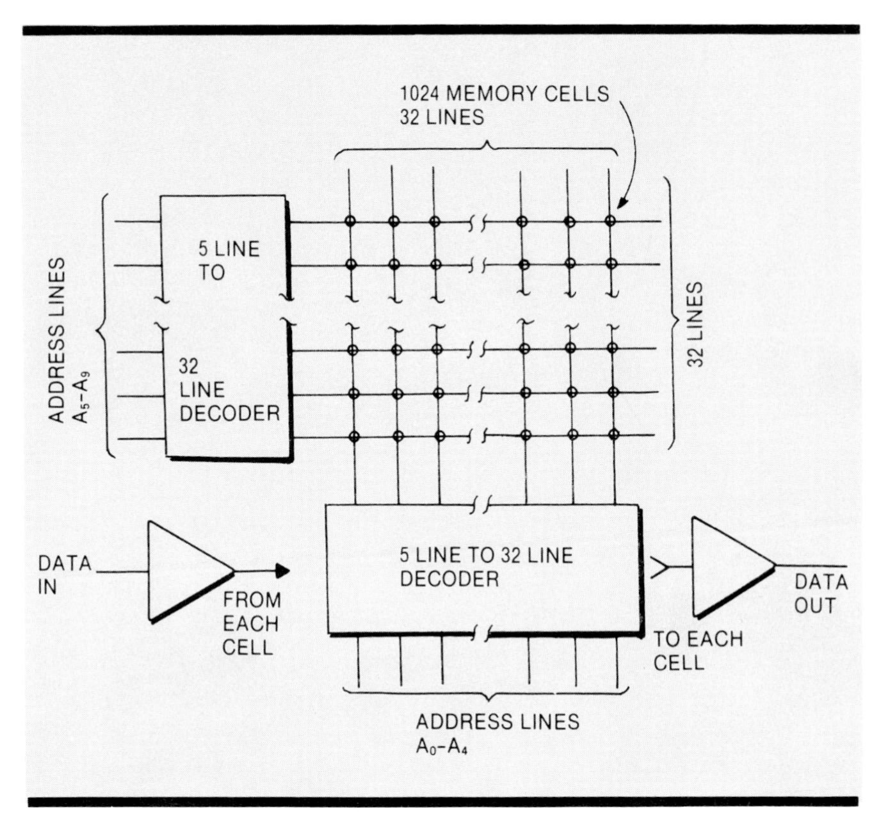

Fig. 5-7. Internal Memory Addressing

together. But the only cell that is fully connected is the particular row-column intersection selected. This means, that to select any individual cell we must only choose between 32 rows plus 32 column connections or two of 64 instead of one of 1024.

This *matrix* organization of connections (in telephone work it is called a *crossbar switch*) results in a savings which is dependent on the row-column organization of the cells. If all cells are in one row or one column, there is no savings—1024 connections are still needed. The optimum arrangement of N cells is a square with each row and column having a length equal to the square root of N. The square root of 1024 is 32, so this is an optimum matrix.

Usually, when a new MOS technology is introduced, the first commercial units are fabricated in a 1×N-bit configuration, but as more confidence in the technique is accumulated, the chips become larger and "wider" and more useful configurations such as N × 4 (nybble) or N × 8 (byte) are produced.

Matrix addressing is a great improvement but is still not enough to satisfy us. Sixty-four pins are far too many; for economy most memory chips are confined to 18 total, so we need another trick. The answer is found in the concept of *decoding*. Look again at Fig. 5-7 and you will see that each of the 32 lines to the rows or columns emits from a "magic box" with only five lines entering. This is labeled a *5-line-to-32-line decoder*. In the next section we will see what is inside this box.

5-4. Decoders

How can the decoder of Fig. 5-7 convert five address lines to 32 separate connections? If you remember the lessons of binary arithmetic, you might better ask "why not?" Recall that we can distinguish 32 decimal numbers with only five binary digits (note that we actually only need two decimal digits). Mathematically, we can say that the number of permutations of five digits, each chosen from the two states, 0 or 1, is 2 to the 5th power or 32. In terms of our decoder, each address line entering can be in one of two states, high or low, so that they can be considered as binary digits with different column weights. Consequently, five address lines can "count" from 00000 to 11111 binary, or 0 to 31 decimal.

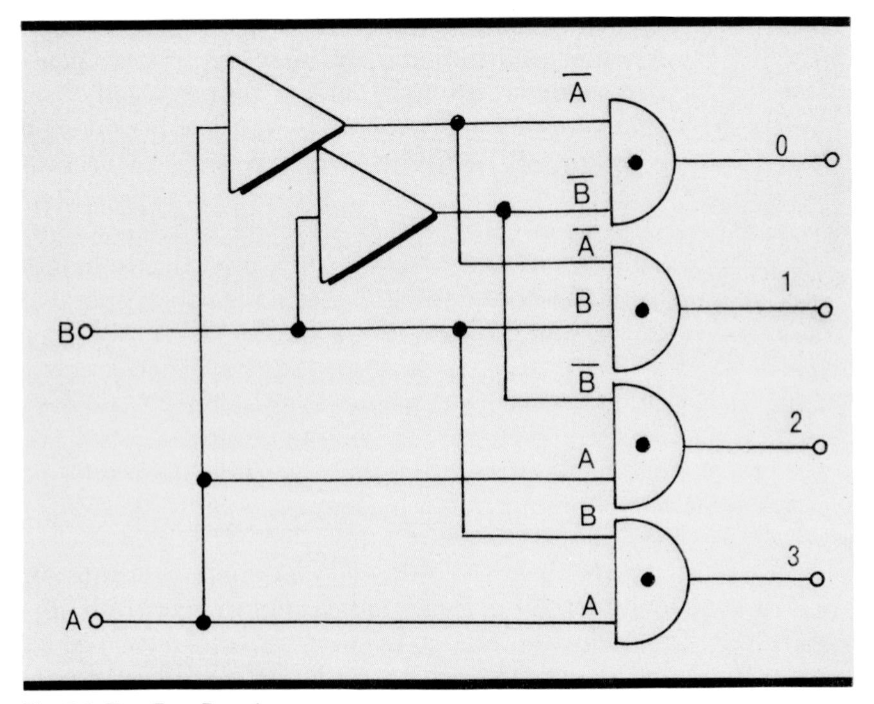

Fig. 5-8. Two-Four Decoder

Figure 5-8 shows how a slightly simpler decoder, "2 to 4," is designed logically. Larger decoders work just the same, except they have more logic elements and more confusing connections. The two input (address) lines can select four outputs. For example, the decimal "1" output is active (high) only if the two inputs to the AND are high, which in turn requires that address line B be high and A false. Table 5-1 is the complete truth table of the 2-to-4 decoder.

Inputs		Output
A	B	Active
0	0	0
0	1	1
1	0	2
1	1	3

Table 5-1. Truth Table 2-to-4 Decoder

5-5. Memory Organization

Decoders are available as separate circuit (IC) chips, for example, the Texas Instruments SN54154 4-to-16 line decoder/demultiplexer. But for the purpose we are discussing they are incorporated into the body of a memory chip. Decoders allow the number of external pins to be kept small and permit the organization of memory chips into various internal configurations to suit different microcomputers and applications. Figure 5-9, for example, shows schematically the arrangement for a 1K×1-bit memory chip. The 1024 cells may be arranged in a 32×32 array as in Fig. 5-7, requiring 10 address lines and 10 pins. Additional pins are needed for data in and data out, read/write select, power and ground, and *chip enable* for a total of 16 pins. The chip enable (also called *chip select*) permits several chips to be tied to the same address lines to be selected (enabled) by higher order address lines or by externally decoded signals. A typical example of this organization is the 2102 static RAM package.

Another memory chip organization is seen in Fig. 5-10. This is a 4K×1-bit chip and requires 12 address lines (6 column and 6 row) to be decoded into a 64×64 matrix. The pinout can be the same as Fig. 5-9, except that 18 pins instead of 16 are required. The Intel 2141 static RAM chip is an example of this organization.

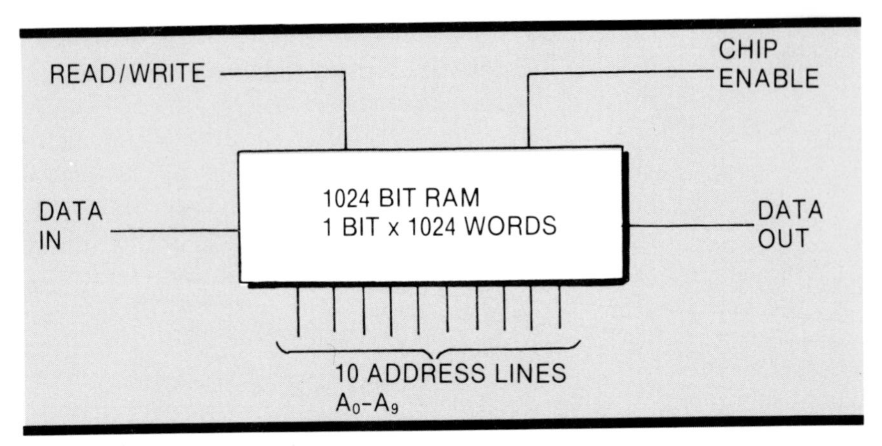

Fig. 5-9. 1K x 1 Bit Memory Chip

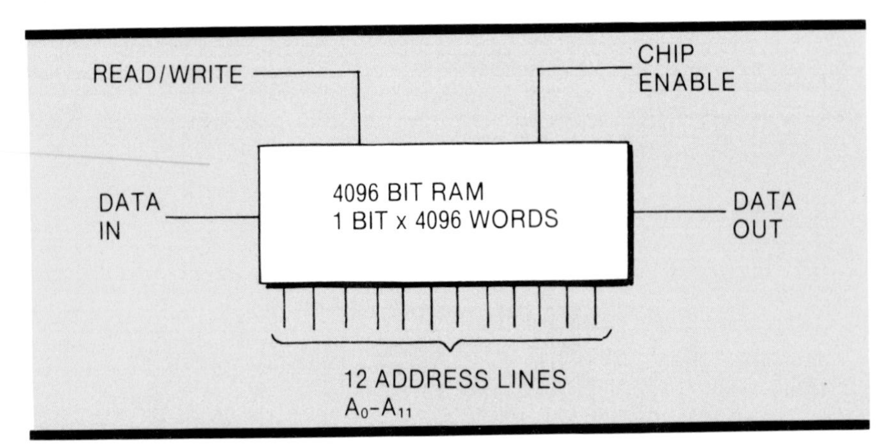

Fig. 5-10. 4K x 1 Bit Memory Chip

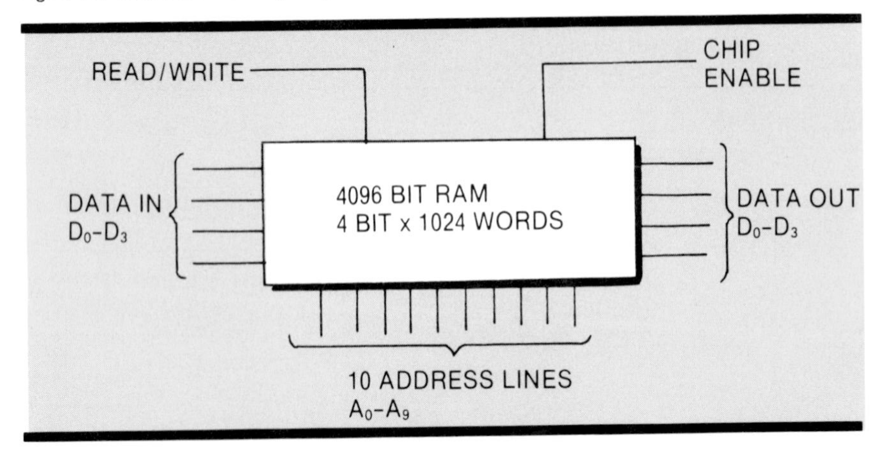

Fig. 5-11. 1K x 4 Bit Memory Chip

Figure 5-11 shows still another arrangement, the 1K×4-bit RAM, or 1024 words of 4 bits each. The 10 address lines are decoded in this chip to select 4 bits at a time, a complete nybble. The pinout can be the same as Fig. 5-9 except, of course, that four common data input/output pins are needed for the simultaneous 4-bit input or output. The popular 2114-type RAM chip is an example of this type.

5-6. Complete Memory Structures

We can see now that it is possible to select a bit or word from a large memory using only a few address lines. In particular, we can address up to 2^{16} or 64K bit words with only 16 address lines. For instance, 1024 bytes could be stored using eight 1K×1 chips in parallel, with 1 bit of each byte residing in a different chip. All 8 chips would be then selected by one chip-enable signal and the 8 bits addressed simultaneously. To add another 1K words of memory would necessitate another 8 chips which would be selected by another chip-enable line.

Suppose we wish to construct a 64K-byte memory for a microprocessor. A more convenient organization than the above would use the 1K×4-bit chip such as the 2114A type used in many systems. Since this chip outputs 4 bits for each address, only 2 chips are needed for each byte. Referring to Fig. 5-12, the data-output lines of each pair connect to the 8 data bus lines, D0 to D7, as shown. Each pair of chips holds 1024 bytes. For a 64K-byte memory we will require 128 chips, occupying at least two average-size printed-circuit boards (PCs) equipped with data bus lines and connectors. (In most real systems, less than the full 64K-byte address space would be occupied by RAM, since much of it will be shared with PROMs, EPROMs, or I/O devices.)

Every chip will be addressed by the same 10 address lines A0 to A9, as shown in the figure. The chip-enable lines of each pair (1K bytes) will be connected, so that both the high and low nybbles are selected simultaneously. Chip selection is achieved by using a 6-to-64 decoder, which utilizes the remainder of the 16 available address lines. These lines carry the 6 higher order bits of the address (A10 to A15).

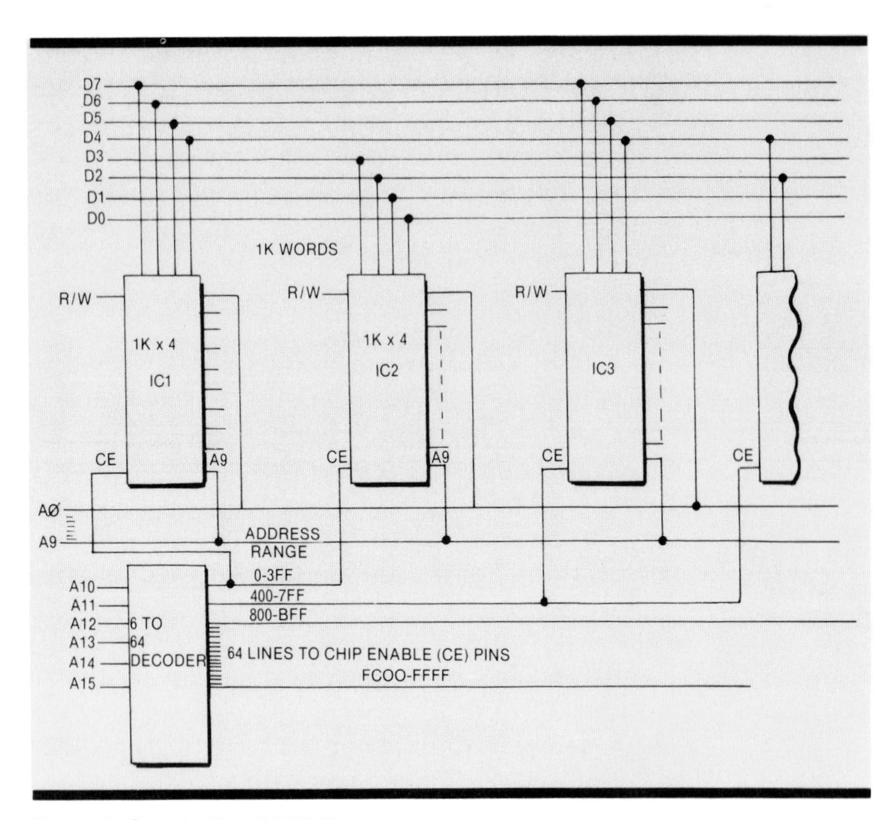

Fig. 5-12. Organization of 64K Memory

The decoder works as follows. Suppose we wish to select one of the 1024 bytes in the lower 1K of address space, that is, hexadecimal 0000 to 3FFF. The inputs to the decoder address lines are:

000000

which selects the chip-enable of the first two RAMs, IC1 and IC2. These chips contain the lowest address:

0000 00 00 0000 0000 (binary)
(high 6) (low 10 bits)

or $0000 (hex)

to the highest address in their range:

0000 00 11 1111 1111 (binary)

or $03FF (hex).

If the highest 6 address bits were:

000010

the decoder would select the third CE line to IC's 5 and 6 (not shown in the figure). These chips contain the lowest address:

0000 10 00 0000 0000
(high) (lowest 10)

or $0800

to the highest address:

0000 10 11 1111 1111

or $0BFF

The 64th chip-enable lines are activated when the higher 6 bits are all ones:

$111111

which permits access to the address range $FC00 to $FFFF, thus completing the 64K-byte address space.

In many systems, a memory board, holding say $4000 (16K bytes) of memory, is constructed so that it can be addressed to any boundary that is a multiple of $1K or $2K. This is done by providing switches or jumper points that can connect the chip-select lines to any desired output of the address decoders.

5-7. Address Space Summary

We have seen how a bus consisting of only 16 lines can address up to 64K words of address space. This space need not be all memory, but can include input/output ports for peripherals, or, indeed, any "addressable" device such as a disk controller. The concept of address space as used here is very broad and useful. For this reason, most microprocessors in common use today are "bus oriented," that is, they carry the concept of address space down to the level of the MPU chip itself. We will return to this idea in the next unit when we consider the *architecture* of the MPU: how its parts interact. However, not all microprocessors are bus oriented. Some use special lines to carry input and output data rather than a common data bus. We will not consider these MPU's in any detail in this text.

Exercises

5-1. *How many AND functions are required to construct a 4-to-16 decoder, using the same logic plan as Fig. 5-8?*

5-2. *How many outputs are possible for a decoder with 10 address lines?*

5-3. *What is the least number of pins required for a 4K×8 static RAM chip?*

5-4. *Match the MOS technology with the following types of processor requirements:*

a) PMOS	*1) Inexpensive, slow calculator*
b) NMOS	*2) Highest speed computer*
c) CMOS	*3) Lowest power consumption*
d) Bipolar	*4) Medium-performance microprocessor or memory*

5-5. *Why is R/WM a better terminology than RAM?*

5-6. *For permanent program storage in the initial models of a new microprocessor controller, what would you choose? RAM, ROM, PROM, or EPROM?*

5-7. *When should one use disk storage rather than magnetic tape?*

5-8. *What is a von Neumann machine? What are its main characteristics?*

Unit 6: Microprocessors

UNIT 6

Microprocessors

In the last unit, you learned something about the technology of MOS LSI chips and how they are fabricated and organized into binary memory structures. The same technologies used to fabricate these relatively simple and regularly organized structures can just as easily produce all the complex circuitry of microprocessors on the same kinds of chips. There is a wide choice of these MPU chips presently available and more are being designed each year. This unit will help you understand the basic features common to most MPUs and their operating principles.

Learning Objectives — When you have completed this unit you should:

A. Know the basic features of an MPU.

B. Understand the architecture of a typical MPU: the major parts and their relationships.

C. Know the function of the main units of the MPU: the ALU, accumulator, program counter, instruction register, decoder, timing, and control.

D. Understand the function of index and general registers, stack pointer, status registers, and flags.

E. Know the fundamental principles of microprocessor timing and cycles.

F. Recognize the concepts of microcoding and microinstructions.

6-1. Microprocessor Features

MPUs are large-scale integrated (LSI) chips, fabricated by MOS techniques in current practice, used for low-cost, high-speed digital data processors. There are a large number of MPU types produced by many manufacturers in the U.S. and abroad. They range in power and sophistication from the simplest — capable of handling calculator arithmetic — to chips that rival the capabilities of large minicomputer CPUs. More capabilities are being introduced into microprocessor chips each year. Some chip sets already have been manufactured that have the capability of an IBM 370 main-frame computer.

Some of the features that determine the capability of a microprocessor unit are:

Data word size

Address word size

Technology (fabrication method)

Power-supply requirements

Packaging

6-2. Data Word Size

We already have learned that word size should bear a relationship to the accuracy and kind of data to be processed. We also can expect that MPUs will be simpler and cheaper, the smaller the word size. The earliest microprocessor, the Intel type 4004, was only 4 bits "wide" since it was developed for an arithmetic hand calculator. Four bits are sufficient to do calculator arithmetic and to display decimal numbers in BCD. The most recent MPUs, on the other hand, such as the Intel 8086, the Motorola MC68000, and the Texas Instruments TMS99000, employ 16-bit data words and can manipulate 32-bit words internally. They will thus support many applications previously employing minicomputers, including the use of high-level languages. Nevertheless, the large majority of MPUs used today have a data word size of 8 bits. Among the most popular are the 6500, 6800, 8080, Z80, and COSMAC families. An 8-bit data word can represent 256 different states, which encompasses all alphabetical data (upper and lower case) and integer arithmetic to a precision of about 1/2%. All microprocessors have the ability to handle words of greater size than their basic bit width for improved arithmetic precision, but this requires performing each operation (such as a fetch from memory) two or more times, and thus costs more in processing time and memory. Some of the 8-bit processors mentioned have, therefore, an additional capability to manipulate 16-bit words, but the data must still be brought in and out of memory in 8-bit bytes.

Another important consequence of data word size is the number of possible instructions. It is an important characteristic of the modern microprocessor CPU that data and instruction words are both transmitted on the same bus and handled in the same way. It follows that the size of the data word and the number of instructions that can be expressed in one word are related. The 256 possible states of the *op code*, the first byte of an instruction, can yield 256 different instructions. Practically, these possiblities are not all utilized, for various reasons. Thus, the 8080A has 78 instructions which are expanded by the use of different address modes (to be explained later) into 121 op codes. The 6502 has only 56 instruction types but 11 addressing modes for a total of 141 combinations used. The Z80 has 158, including the 78 of the 8080A, so that the programs written for the latter can be used. In contrast, the 4004 would seem to be limited by its 4-bit word to 16 instructions, a not very useful device, but in actuality it employed an 8-bit instruction register and so was able to achieve 46 separate op codes. The Z80 also extends its instruction capability by using 2 bytes in some op codes, which, of course, require two loadings from memory; the more common codes use 1 byte. At the other extreme, the 16-bit MC68000 combines 56 instruction types and 14 addressing modes to over 1000 separate instructions.

6-3. Address Word

The size of the address word, as we saw in Unit 5, determines the number of memory locations directly addressed. The usual address word and bus is 16 bits wide for a capability of 65,536 locations. The MC68000 has a 24-bit bus for over 16 million locations. Memory space can be extended beyond that of direct addressing by switching in banks of memory. This trick was used to extend the 4004's range. Other schemes used to extend addressing range depend on special features of the processor or instruction set. These are found in the more expensive personal computers but are seldom justified for industrial controls. If a large memory is needed, it would be advisable to consider a 16-bit processor.

6-4. MPU Technologies

Technology features already discussed apply to microprocessors as well as memories. PMOS is normally confined to the 4-bit calculator applications where its very low cost is an asset. NMOS is the most common for 8-bit and larger devices, owing to the higher density of gates and complexity possible, together with higher speed and modest power needs. CMOS is used for very low-power applications, as in cases where long-period battery operation is required. Its superior noise immunity is also an asset in some applications. Bipolar, with its very high speed and low density characteristics, is usually confined to multiple chip (*bit slice*) applications in which a number of 2-bit or 4-bit MPUs are joined by common buses to form a larger word unit. Bipolar bit-slice architecture is usually confined to minicomputer designs.

6-5. Power Supplies for MPUs

Since microprocessors distinguish between 0 and 1 bits at the level of a few volts and since volatile memories lose stored data on power loss, the regulation and stability of power supplies is important. A good, regulated power supply may cost much more than the MPU chip. All other things being equal, the fewer chip power requirements lead to the most reliable and least costly system.

The 8080A microprocessor requires three voltage levels: +5, −5, and +12 volts. A more recent version, the 8085A, operates on a single 5-volt supply. This is also true of the Z80 and 6502 processors, all using NMOS technology. CMOS processors such as the RCA COSMAC operate with a wider range of power inputs, between 4 and 10.5 volts, contributing to their noise immunity.

6-6. Packaging

The packaging feature of most concern to the MPU designer and user is the pinout: the number of connections or pins required to interface with the world external to the chip and their assignments. The package is usually a dual in-line (DIP) encapsulation of the chip, similar to Fig. 6-1. Packaging other than DIP is sometimes used for military and high-temperature use, but the principle of pinout limitation remains the same.

The pinout is important for two reasons: the number of contact leads brought out of the chip affect the cost directly and the chip reliability inversely; and, to standardize on socket manufacture and printed circuit board automated assembly, it is desirable to limit DIPs to standard sizes. For example, 16 pins is the standard for conventional logic ICs and 18 for most memory chips. The

8-bit microprocessor with full addressing capability is usually packaged in a 40-pin DIP.

The limitation of 40 pins means that some functional compromise may be necessary. The difference between 8-bit microprocessors then depends as much on how designers choose to utilize the available pins as it does on state-of-art technology and price objectives.

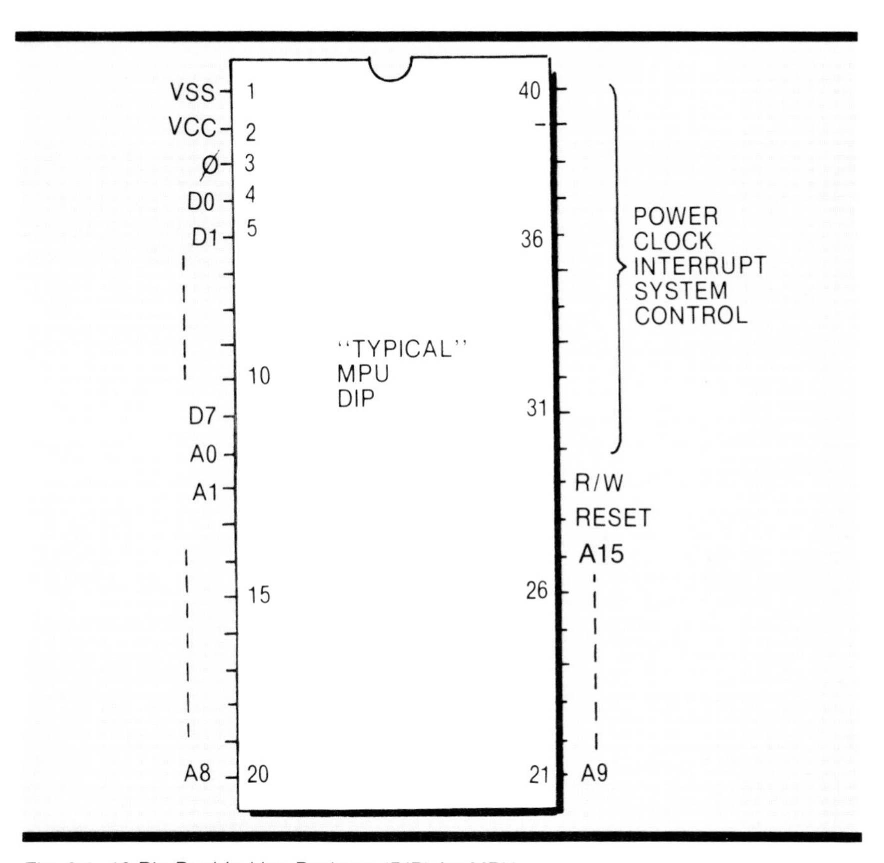

Fig. 6-1. 40 Pin Dual In-Line Package (DIP) for MPU

Figure 6-1 demonstrates the necessary allocation of pins for a "typical" MPU package. Obviously, the 8-bit machine must have eight data lines; these are shown as D0 to D7. Likewise, for full addressing capability there must be 16 address lines, A0 to A15. At a minimum, there must be one pin for a single power voltage (Vcc) and one for ground (Vss). In addition, there must be at least one clock signal in O, one RESET to initialize operations, and at least one READ/WRITE (R/W) signal to control data transfers between MPU and memory. Thus, the minimum number of pins is 29, and the 11 remaining are available for "refinements."

Table 6-1 is a comparison of the pinouts in the 6502, 8080A, and Z80 microprocessors.

	6502	8080A	Z80
Data bus	8	8	8
Address bus	16	16	16
Power	1	3	1
Ground	1	1	1
Clock in	1	2	1
Clock out	2	—	—
Read/write	1	2	4
Reset	1	1	1
Interrupt	2	2	3
System Control	2	3	2
DMA	—	2	2
Overflow	1	—	—
Dynamic memory	—	—	1
GND./NC	4	—	—
Totals	40	40	40

Table 6-1. Pinout for Three 8-Bit Microprocessors

Some of the individual features of these MPUs can be noted from the table:

The 6502 has a single 5v power requirement and also has an "on-board" clock for synchronizing operations. This means that only a single clock frequency from a crystal oscillator (external to the MPU) need be supplied; the MPU generates two phased square waves for the timing operations. Two additional pins are used to output these two phases so that devices external to the MPU, such as peripheral interface chips, can be synchronized with the processor. The 6502 has two levels of *interrupt* signal: interrupt is a technique for servicing an external device, such as a printer or a keyboard, which is not scanned by a program. It also has two other signals, SYNCH and READY used for system control; these interface the CPU with slow external memories and also permit single-step operation of the processor. The 6502 also has an OVERFLOW pin to allow an outside signal to set the overflow flag (explained in Sec. 6-10). These total 36 pins; of the four left over, one is an extra ground and the other three are not connected.

The 8080A requires three pins for power: +5v, −5v, and +12v. Since it has no on-board clock, it requires two pins to enter the two phases. Writing and reading from memory demands two pins; RESET and interrupt functions require three, and three more for system control.

The two pins remaining of 40 are used to implement a function called *direct memory address* or DMA. DMA is a hardware method by which a fast peripheral may exchange data with RAM without going through a processor program, which might slow up the transfer. In one technique, known as *burst DMA*, the microprocessor operations are halted and the MPU connections to the address and data bus are forced into a high impedance state by means of a circuit called a *three state or tri-state*. The effect of this circuit is to disconnect the MPU from the buses and allow them to be taken over by the external device, which will then read from or write to memory. Another DMA technique called *cycle stealing*; this allows processing to continue but "steals" the bus during portions of the operating cycle not used by the MPU. DMA can be accomplished in the 6502 only with the aid of external tri-state circuitry. (However, another member of this family, the 6508, does have a tri-state address buffer.)

The Z80 pinout is listed in the third column of Table 6-1. Like the 6502 it has a single power voltage and needs only one clock input. A more elaborate R/W control requires four pins. Three more are associated with interrupt and two (WAIT and SYNCH) provide system control as with the 6502. Like the 8080A, the Z80 has two DMA signals, Bus Request and Bus Acknowledge, controlling the address and data bus tri-states. The remaining pin implements dynamic memory refresh by signaling the presence of a memory address requiring refresh by a read operation.

6-7. MPU Architecture and Operations

Before discussing architecture, that is, the organization of the various units within the MPU, we first should define the *register*, a fundamental building block of the processor. A register is merely a place to store a word of binary data. It is distinguished from main memory because it need not be addressed. A register may consist of a number of bit cells, like 8 or 16, each cell being connected to one line of a bus. Movement of data into a register (assuming that it already exists on the bus) merely involves gating or enabling the register cells to read it. Conversely, a register may write to a bus by being enabled in the opposite direction. The principal MPU registers (but not all) are bidirectional. A *buffer* is much like a register, but is usually meant to store data while awaiting some timing event or strobe; consequently data moves through a buffer in one direction. The movement of data in and out of registers and buffers is controlled though segments of the MPU instruction referred to as *microinstruction*. These we will consider when we examine the inner workings of the MPU.

Figure 6-2 shows how the registers are interconnected with the ALU and other MPU sub-units. The important elements of this architecture are:

ALU

Accumulator register

Program counter register

Instruction register

Decoder and timing unit

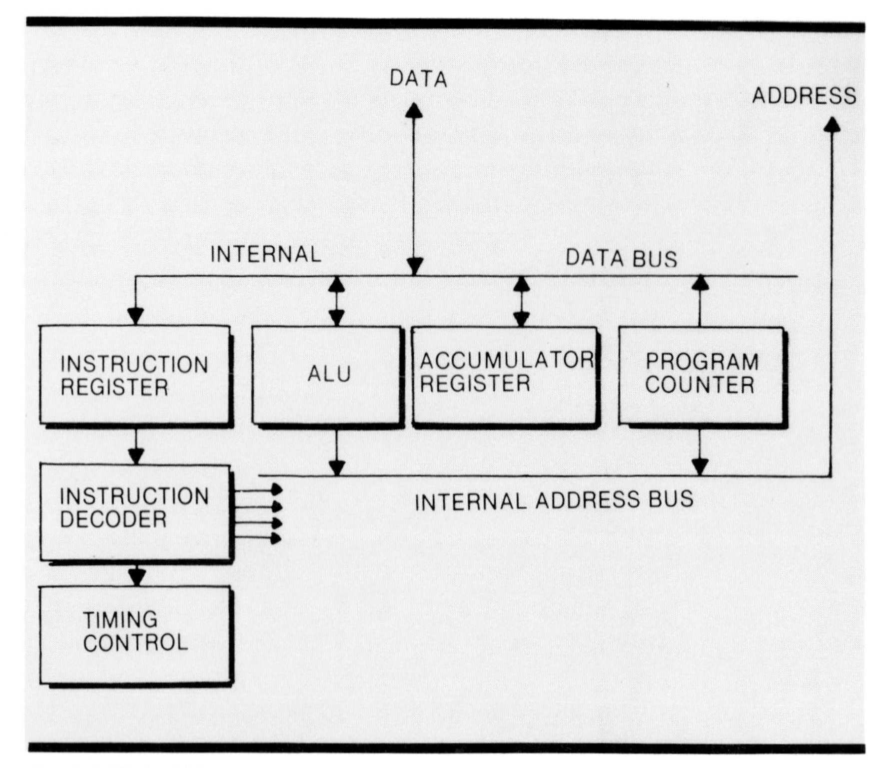

Fig. 6-2. Typical Microprocessor Architecture

The ALU has already been described as the device in which addition, complementing, logical operations, and shifting or rotation of word bits are carried on. The *accumulator register* (A) works in partnership with the ALU by accepting bus data for the ALU's operations and receiving its results so that they can be put back on the bus. Since the ALU often requires two inputs for logic or arithmetic it also is capable of receiving data directly, or through a buffer connected to the bus.

The *program counter* is a register of fundamental importance. It holds the address of the next word to be fetched by the MPU. To connect with the 16-bit address bus it must be 16 bits wide, or more commonly, consist of two 8-bit registers, one to hold the most significant part of the address (hi byte) and the other to hold the least significant 8 bits (lo byte). The important feature of the PC is that it acts as a counter, meaning that it automatically increments (increases its count by one) after each fetch. In the normal course of events, after completing an operation specified by an op code previously fetched, the program counter will fetch the op code held by the next sequential address in program memory. This very simple scheme works because the computer's program is invariably written to an ascending numerical sequence of addresses. In other words, it is not necessary to build in an elaborate scheme to find where the next instruction is located; by default it is found at the next address.

However, the contents of the program counter can be changed by an arbitrary amount by the program itself, using codes known as *jump or branch* instructions. These will be discussed at more length in succeeding units on programming.

The *instruction register* receives the op code and stores it until the instruction is completed. As shown in the figure, this register is directly connected to the *instruction decoder*. This unit interprets the op code and, together with the *timing control*, issues detailed instructions to the ALU and to other registers. The instruction decoder has been called the "CPU of the microprocessor," meaning that it acts like a small computer in its own right. The 8-bit op code is broken down by the decoder to generate a small program of consecutive actions that move and manipulate data through the ALU and registers. These actions are called *microinstructions* and the program a microprogram. In some CPUs the microprogram can be altered by the programmer, that is, the CPU is *microprogrammable*. Generally, microprogrammable microprocessors are of the "bit slice" variety, 2-bit or 4-bit wide units that can be assembled to a single wider-word unit, usually part of the architecture of a minicomputer. We will not consider this architecture further here. In the common types of microprocessors we are now discussing, the microprogram of the control unit is a fixed logic established by the chip designer and implemented in ROM or "hard-wired," thus not accessible to the programmer. In fact, the microprogram is of very little practical interest to the programmer who is really only concerned with the macroresults of the instruction and the type and number of instructions available in the microprocessor's instruction set.

However, the microinstructions for each microprogram must occur within a time frame compatible with the overall *instruction timing* of the MPU. To understand this, we must realize that every MPU instruction follows a sequence of operations timed by a crystal clock which generates *clock phases* driving the timing control unit.

Every instruction of an MPU consists of two stages: *instruction fetch* and *instruction execute.* The following sequence is common to all instructions:

1. The value of the program counter register is gated to the address bus.

2. A signal (strobe) causes the instruction to be read from memory into the instruction register, completing the fetch.

3. The op code in the instruction register initiates the timing control and decoding microprogram.

4. The program counter is incremented by one to ready it for the next operation.

The remainder of the instruction execution depends uniquely on its type and nature.

The timing of the instruction fetch is detailed in Fig. 6-3. The upper diagram shows the clock timing waveform. (There may be more than one of these waveforms in some microprocessors, bearing a phased relationship to each other.) At a time determined by the rise or fall of one of these clock phases, the address bits are gated on to the address bus. Since timing is not perfect in the gating mechanism, there is a small interval before all the address bits are in their stable positions and are gated by the strobe to the bus. During this period, the READ signal causes the external logic of the selected memory chip to put the data value on the data bus. When the data is stable, another strobe gates the data on to the instruction register.

All operations require instruction fetch and thus at least two clock cycles are needed for an instruction: one for fetch and at least one for execute. If data are to be fetched also, another cycle, identical to Fig. 6-3, will be required.

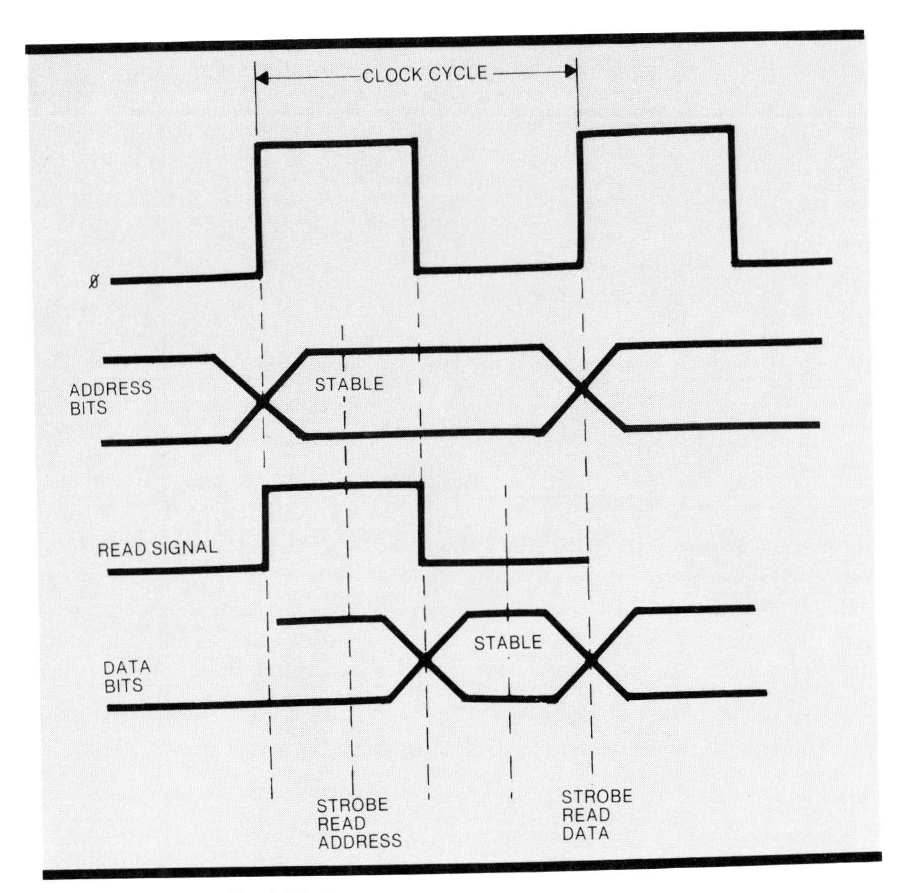

Fig. 6-3. Instruction Fetch Timing

Returning to the action of the instruction decoder, it is necessary that all of the microinstructions making up the instruction-fetch portion of the microprogram be completed within a single clock period. This is required so that the instruction can be decoded during the next clock cycle which also may be used to fetch the next increment of data or address. (This multiple activity during one clock cycle is termed *pipelining.*) The microprogram for instruction fetch may include such microinstructions as: move program counter to address register; set READ control signal TRUE; move PC low-order byte to data bus; and so forth. Depending on the microprocessor, of course, perhaps 16 separate microinstructions are needed to describe all of the instruction-fetch activities. If the clock period is one microsecond, the time allocated for each microinstruction must not exceed 1/16 microseconds or 62.5 nanoseconds. This is a time period compatible with the MOS construction of MPUs.

6-8. Index and General-Purpose Registers

The MPU described in Fig. 6-2 is a "bare bones" device and contains only the minimum parts necessary to be called an MPU. Real-world microprocessors have many more types of registers as well as duplicated units, so as to be able to give the greatest flexibility and freedom to the programmer, to increase the instruction set, and to obtain better execution speed. *General registers,* you will recall, are those that can be addressed by the programmer, that is, they can be loaded with data and their contents transferred by instructions. *Index registers* are a particularly useful type of general register because they facilitate the generation of addresses used in programs. For example, if it is necessary to address a sequence of memory locations starting at some arbitrary point, the initial address of the block may be loaded into the accumulator and an instruction given to add this base address to the number held in the index register. This number can be increased automatically until the entire block of addresses has been read. *Indexed addressing* thus adds considerable flexibility to programming since the base address can be easily changed to read another block, etc. Index registers, used as counters, also can be employed to repeat sections of a program for a given number of times; this is known as *looping* and is an important programming trick. If, for example, we would like to print the numbers 1 through 8, we can accomplish this by writing a routine to print the number 1, using the ASCII code, and then repeat this routine, adding one to the ASCII value at each repetition until we are through. The index registers could be used to increment both the ASCII value, which determines the number printed, and the count of the repetitions.

The 6500 series microprocessors have two index registers, X and Y, both being 8-bits wide. The 8080A has six 8-bit general-purpose registers acting in pairs; they can be used as

16-bit registers and thus add more flexibility to addressing or arithmetic. The Z80 goes further in having two sets of six general-purpose registers that may be used individually or in 16-bit pairs. There are also two sets of accumulators. The programmer may access either set of registers through exchange instructions, allowing two sets of programs to be running alternately on a "foreground-background" basis.

6-9. Stack and Stack-Point Registers

The *stack* is another programming convenience supported by most microprocessors. It is merely a dedicated block of contiguous memory locations that are used for temporary storage of data. According to D. Knuth, a stack can be defined as a linear list of addresses in which all insertions and deletions of data are made from one end of the list. During an *interrupt,* for example, when the microprocessor exits from a program and services the interrupt routine, it is necessary to store the contents of the program counter and working registers so that it will be able to return to the main program and take off from the point where it was interrupted. The stack is a convenient place to store this information and from which to retrieve it when needed (since it will be on top).

Another example of stack utilization is with *Reverse Polish Notation (RPN)* or *post-fix* arithmetic. This notation is familiar to users of Hewlett-Packard calculators. RPN eliminates parentheses by loading data and operations on a stack, like trays on a cafeteria dispenser, where the last item placed on top of the stack is the first removed. This is called *last-in first-out (LIFO).* The mathematical operators will act on the one or more data items on top of the stack, taking as many as needed. For example, let the stack hold the numbers:

BOTTOM> 5 15 22< TOP

(where BOTTOM and TOP refer to the stack). The the operator + in RPN will add the top two items and return the value 37. The advantage of stack of RPN notation is that the storage implied by parentheses is not required in complex operations such as:

(A + B) * (C + D)

The high-level programming language FORTH uses RPN and stack operations, as we will discuss in the next unit.

The *stack-point register* or stack pointer is merely a counter that keeps track of the address called the "top of the stack." The stack pointer is always left pointing to the next memory address in which data can be stored. In the 6502 and many other MPUs the stack is "built down," that is, the top of the stack is initially the highest address of the stack and as data items are added they occupy progressively lower addresses. The 6502 processor uses effectively a 9-bit register for its stack pointer with the most significant bit permanently set to 1, therefore only the addresses $0100 to $01FF (page one) are available as the stack area. (Note that 256 contiguous locations in memory are called a *page*; the page number is the two high-order hex digits of the address. Thus, $0100 to $01FF is page one, which is reserved for the stack in the 6502. Locations $0000 to $00FF, page zero, have also a special connotation in the 6502 which we will consider in the next unit.)

Example 6-1

Figure 6-4 demonstrates the use of the stack in the 6502 microprocessor. The instruction mnemonic PHA, meaning "Push Accumulator on Stack" stores ("pushes") the accumulator contents on to the top of the stack in the next available location. In the example (a) the accumulator register (A) holds the byte $EF and the stack point register (SP) holds the address $1A, translating to a stack location $011A. When the PHA instruction (actually the op code $48) is read by the instruction register, $EF is stored in $011A. The value in SP is automatically decremented by one to $19, see (b).

At some later time, it is desirable to restore ("pull" or "pop") $EF from the stack to the accumulator. The instruction PLA ($68) will increment the value of SP and then transfer the contents of the corresponding page one memory to the accumulator. If SP was $19 before the instruction it will increment to $1A and the $EF value in $011A will be restored. (Note that it is the responsibility of the programmer to see that the correct value of SP is present before popping the stack. By means of other instructions, any 8-bit number can be moved into SP, say, from the index register X.)

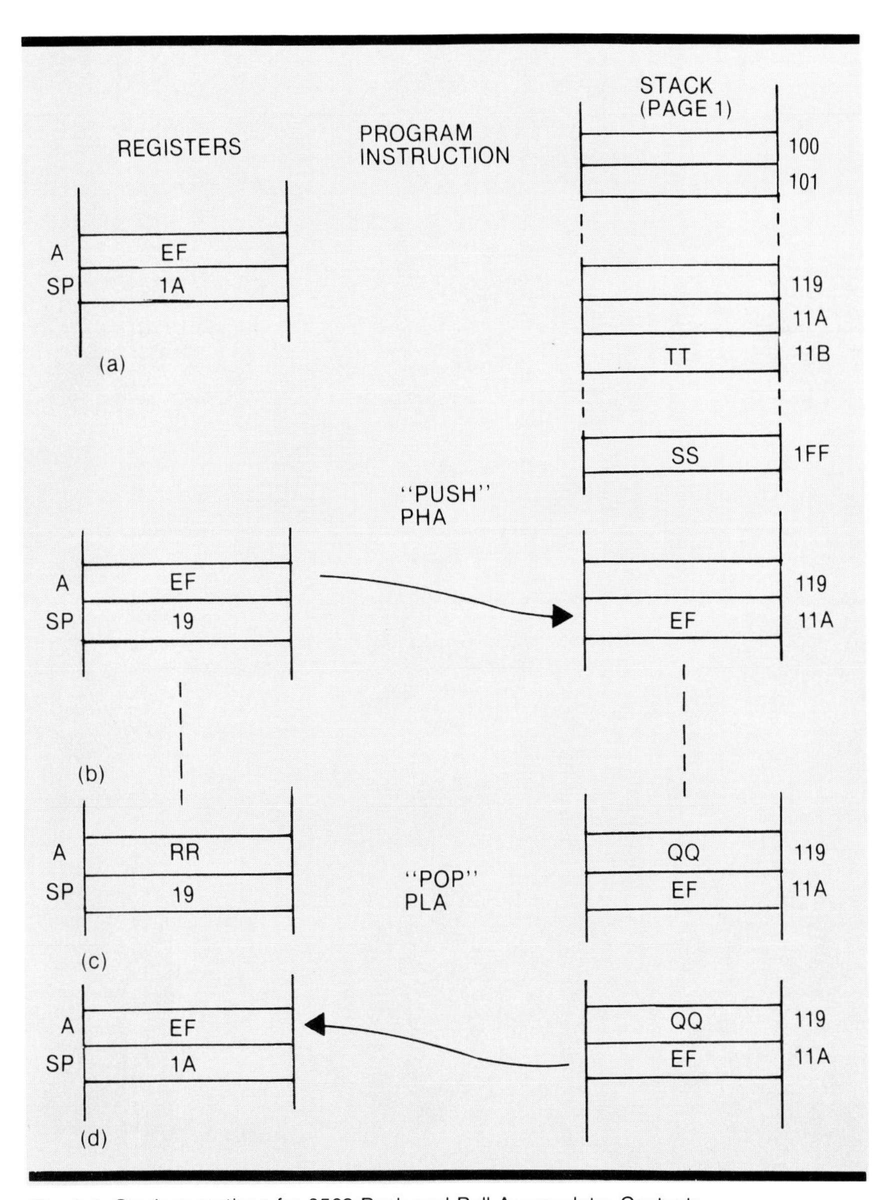

Fig. 6-4. Stack operations for 6502 Push-and-Pull Accumulator Contents

Probably the most important use of the stack is to store the value of the program counter when calling a *subroutine*. Subroutines are segments of code that are used repetitively in a program to accomplish a repetitive task. A subroutine can be called by a specific instruction, such as JSR (op code 20) in the 6502, which is a mnemonic for Jump to new location Saving Return address. The *jump* transfers program control to the address of the subroutine by forcing this address onto the program counter. The program then continues at this new address until the subroutine is completed and an RTS (Return from Subroutine, op code 60) is encountered. The JSR puts the 2 bytes of the next step of the uncompleted main program onto the stack, while the RTS retrieves that address from the stack and puts it back into the program counter, so that the main program continues where it left off. The full value of the stack is seen when it is realized that a subroutine can call another subroutine and so on, through many layers of depth, Fig. 6-5. This recursive process of subroutine calling subroutine is called *nesting*. The 256-byte length of the 6502 stack can nest subroutines to a depth of 128 or can interrupt 85 times without using up the stack area. The use of subroutines saves much space in coding programs so it is widely practiced; however, it is unlikely that nesting subroutines will use exceed the capacity of the 6502 stack.

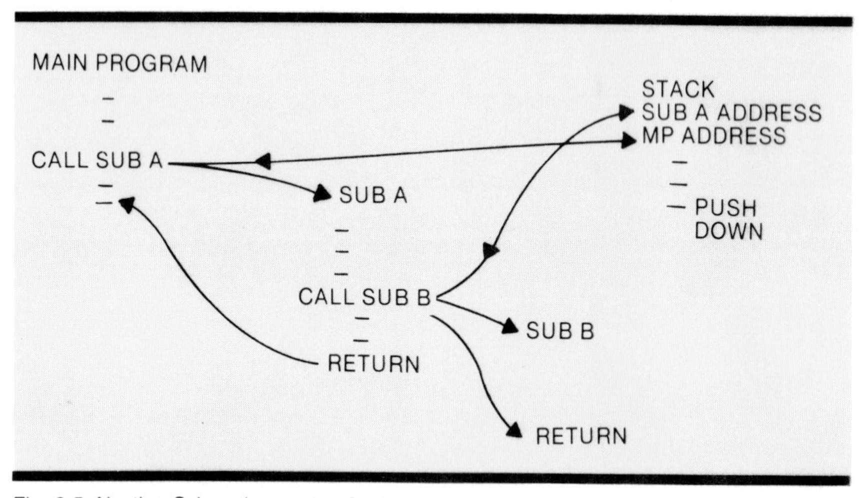

Fig. 6-5. Nesting Subroutines using Stack

6-10. Status Registers

The *processor status register* (P) is the last we will consider in this unit. It also has been called the *flag register* or *flag flip-flops* in some MPUs, which carries the implication that the individual bits of the register are markers or "flags" carrying the important information. The register content as a data word or a counter number is not meaningful. The status register is 8 bits wide in the 6502 but only seven of these bits are used, Fig. 6-6. In the 8080A, five flag flip-flops are provided. The Z80 has several 8-bit registers that can be utilized for flags.

The status flags indicate the result of the previous operation carried on by the processor and thus permit subsequent action — specifically, program branching — to take place, depending on that result. This property is what provides the processor with its "decision-making" capability. Therefore, the most important function of the status register is to allow instructions to be designed that implement a program jump or branch *(conditioned branch)* when these conditions are encountered. The conditions noted by the 6502 are shown in Fig. 6-6. They are:

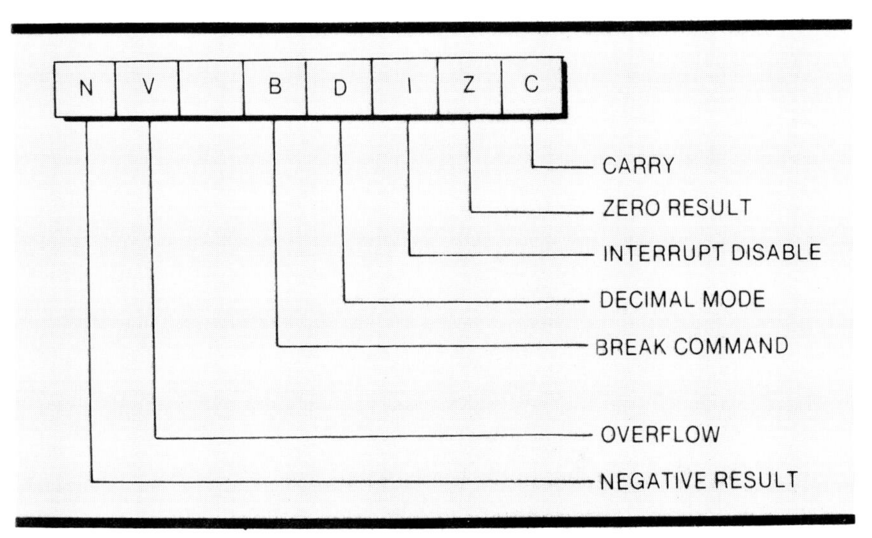

Fig. 6-6. Status Register for 6502 Processor

N — Negative result. Repeats the sign bit (bit 7) of the previous move or arithmetic operation.

V — Overflow. Reveals overflow of signed number arithmetic from bit 6 to the sign bit (bit 7). Is 1 when overflow.

B — Break command. Distinguishes between an interrupt and a programmed break (BRK). Set when interrupted by BRK.

D — Decimal mode flag. The 6502 is capable of performing packed (4-bit) binary coded decimal arithmetic when D set.

I — Interrupt disable. If set to 1 by programmer, signals to the interrupt pin are disregarded.

Z — Zero result. Is set when previous result was zero.

C — Carry. Set when last operation resulted in a carry bit 1.

As an example, if two equal numbers are subtracted the result is 0. Thus, the Z flag is set. This signifies a successful comparison between the 2 bytes. If the conditional branch instruction BEQ (Branch if Equal to zero) is next encountered, the branch would be taken, since Z=1, and the program control would be moved to the new location called for by the instruction by altering the PC value.

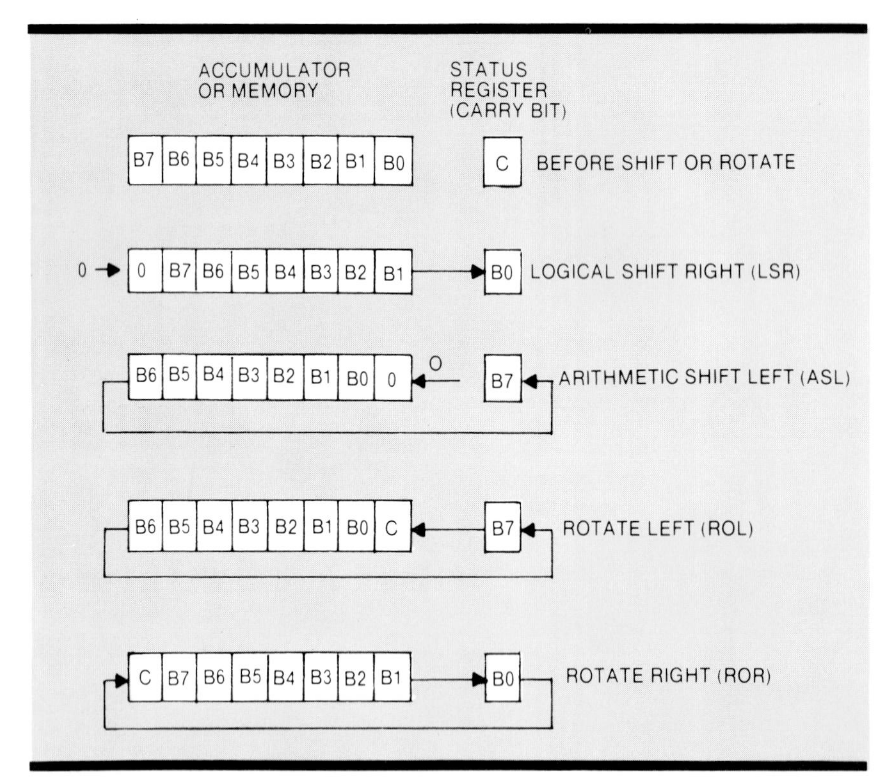

Fig. 6-7. 6502 MPU Shift and Rotate

Likewise, another flag could be used for a branch. The BCC instruction calls for a Branch if Carry Clear (C=0). The carry bit has many important functions, as we noted in the previous units on binary arithmetic. The carry result is needed for multibyte (precision) arithmetic. It also is important in *shift* operations, where the memory or register contents are moved right or left, thus effectively moving the binary point and multiplying or dividing by 2. The bit on the end of the register that is moved out goes to the carry-bit position of the status register. In the 6502 there are several of these instructions, called *rotate* (right or left), and *logical* or *arithmetic shift*, Fig. 6-7. The rotate right (ROR) or left (ROL) move all the accumulator or memory bits one bit in the indicated direction. The bit that "falls off" goes to the carry position (C) of the status register and whatever was in the carry position goes into the opposite end of the shifted register. The arithmetic shift left (ASL) moves the MSB into the carry and a 0 always goes into the LSB of the register. Logical shift right (LSR) works in the opposite sense, putting the LSB into C of the status register and moving 0 into the MSB.

As stated, the 1 flag, called the *interrupt mask bit*, makes it possible to disable or "mask" the interrupt request IRQ. If the programmer has set the flag to 1, the interrupt request is not serviced. (However, recall that the 6502 has a nonmaskable interrupt NMI that will interrupt regardless of the flag setting. Thus there are two levels of interrupt priority in this MPU.)

The D or decimal flag allows the programmer to set the arithmetic mode of the processor. If set, the BCD mode of arithmetic would be carried out, as in the following example.

Example 6-2
BCD addition

0111	1001	79	
0001	0100	+14	(Decimal add)
1001	0011	93	(BCD result)

The B or break flag is set only by the processor. Thus it determines if the interrupt was a "real" (hardware) interrupt, caused by a signal on the IRQ or NMI pin, or caused by a BRK command in the program. The BRK is normally used in debugging programs to hold the action at a particular spot and vector it to a new location where the interrupts can be serviced. If the B bit is set, the processor knows that the interrupt was caused by the BRK and was not "real." Therefore, the program can be written to take the appropriate action.

The use of the V and N bits were listed above. The V bit, showing overflow of a signed number from its MSB (bit 6) to the sign area (bit 7), signifies that appropriate action must be taken by the program. The V bit works like the carry bit when using signed arithmetic, but can be ignored if ordinary binary arithmetic is being used. The V bit can also be used with peripheral chips that use bit 6 for special control signals. The N or negative flag, in addition to its repeat of the sign bit, is useful for many tests and branching instructions. Since it is the highest bit, it can be easily tested or isolated through use of the ASl and ROL instructions.

Exercises

6-1. *What size microprocessor (data word width) would you choose for the following tasks (give an example of each):*

a) Simple arithmetic calculations output to an 8-segment LED?

b) Write reports in English (ASCII code) on a printer?

c) Perform precision arithmetic rapidly, using a high-level programming language?

6-2. *Can the MC68000 MPU address more memory than the 6502? Is there any effect on the number of signal lines entering the chip?*

6-3. *How many op codes are possible with a 16-bit microprocessor?*

6-4. *How is it possible for an 8-bit microprocessor to store and retrieve 16-bit-wide addresses? What is the penalty?*

6-5. *Would you consider CMOS microprocessors for a process-control application? Why or why not?*

6-6. *Give three reasons for using a 40-pin DIP package for MPUs.*

6-7. *Can a 40-pin package be used for a 16-bit MPU? Give reasons.*

6-8. *Assume you are designing an 8-bit MPU for toys or small control functions and wish to save space and cost by using a 28-pin DIP. What function(s) would you reduce or eliminate?*

6-9. *What single register do you think is the most important with respect to communication with memory?*

6-10. *The key feature of the stored program (von Neumann) computer architecture is that it permits data and instruction to the intermixed in memory. What register facilitates this capability? What does it do?*

6-11. *What is the importance of microcode to you if you are a typical MPU user or programmer? Should you learn it?*

6-12. *Should you learn op codes? Why?*

6-13. *What are the two parts of every MPU instruction implementation?*

6-14. *How many clock cycles are needed for an instruction?*

6-15. *What advantage do the register pairs in the 8080A impart when performing arithmetic?*

6-16. *List three applications of indexing.*

6-17. *Give two common examples of a LIFO stack.*

6-18. *What is the advantage of using RPN arithmetic?*

6-19. *How does the stack pointer facilitate nesting subroutines?*

6-20. *What would be the result of Example 6-2 if D=0?*

6-21. *How can the 6502 status register be used as a counter?*

6-22. *What is the main purpose of the status register?*

6-23. *What is the difference between rotate and shift instructions?*

6-24. *In writing a program, you wish to save the sign of a signed number in the accumulator, perform some logical operations, and then restore the sign. What is an easy way to do this? (Hint: it can be done in two instructions.)*

References

Bibbero, R.J., *Microprocessors in Instruments and Control*, New York: John Wiley & Sons (1977), Chapter 4.
Leventhal, L.A., *6502 Assembly Language Progamming*, Berkeley, California: Osborne/McGraw Hill (1979).
Bibbero, R.J. and Stern, D.A.,*Microprocessor Systems, Interfacing and Applications*, New York: John Wiley & Sons, (1982), Chapter 2.

Unit 7: Programming a Microprocessor

Unit 7

Programming a Microprocessor

In the previous unit you were introduced to most of the hardware features and architectural concepts of microprocessor units. An MPU is a data processing machine of boundless versatility waiting for your instructions. In this unit you will start learning how to give these instructions and bend the microprocessor to your will!

Learning Objectives — When you have completed this unit you should:

A. **Understand the use of algorithms and block diagrams in designing software.**

B. **Know the difference between assembly code and high-level languages and where they are used.**

C. **Recognize the names and features of high-level programming language.**

D. **Know how to interpret an MPU assembly language mnemonic and its addressing modes.**

7-1. Programs, Flow Diagrams, and Algorithms

Computer science is divided into hardware and software disciplines, or according to some, hardware, software, and firmware. Hardware is the unchanging core of electronic circuitry, registers, adders, shifters, and bus conductors which we have been discussing. Software is the lists of instructions that cause all of this hardware to work together to produce a useful result, such as adding 2+2 or your income tax, or controlling the steam flow in your plant's boiler. *Firmware* is only technically different from software; it is a software program that has been put in ROM because it will never (hardly ever) be changed. With the increasing use of EPROM and EEPROM, the distinction between software and firmware is becoming blurred.

Software has been called the intelligence of the computer. Indeed, programs are becoming so sophisticated that in fields like chess-playing machines, the term *artificial intelligence* has become popular, meaning programs that exhibit complex decision-making capability and adaptability to the human user, and perhaps, someday, creative ability. Software is also a "buzzword" today for another reason. The number of people with the ability to create the programs needed in industry and elsewhere fall short of the number available. Consequently, programming is expensive, far exceeding the cost of hardware in most cases. It pays to be able to do some programming yourself if you are going to use MPUs as an engineer or technician, and to know enough about it to follow the work of programming professionals when they are called in to perform large or complex jobs.

John von Neumann, the originator of today's computer architecture, defined a program as the replacement of the complex operations of solving mathematical problems by "a combination of the basic operations of the machine."

"Basic operations" are indeed the nitty-gritty of any computer program. Ultimately, the machine "understands" only the sequence of ones and zeros put on its bus lines so they may be decoded into microinstructions and addresses.

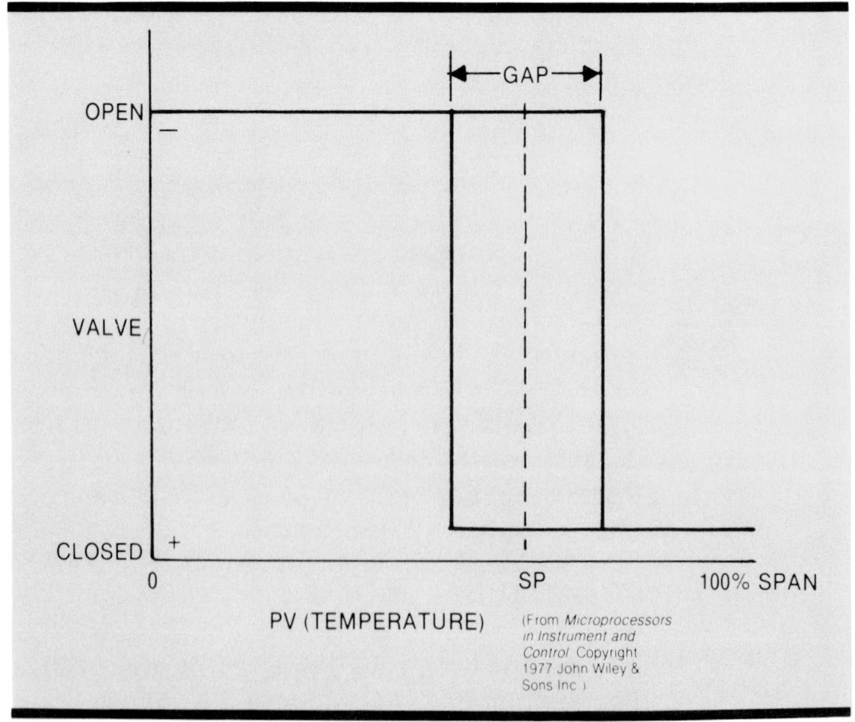

Fig. 7-1. On-Off Control Mode

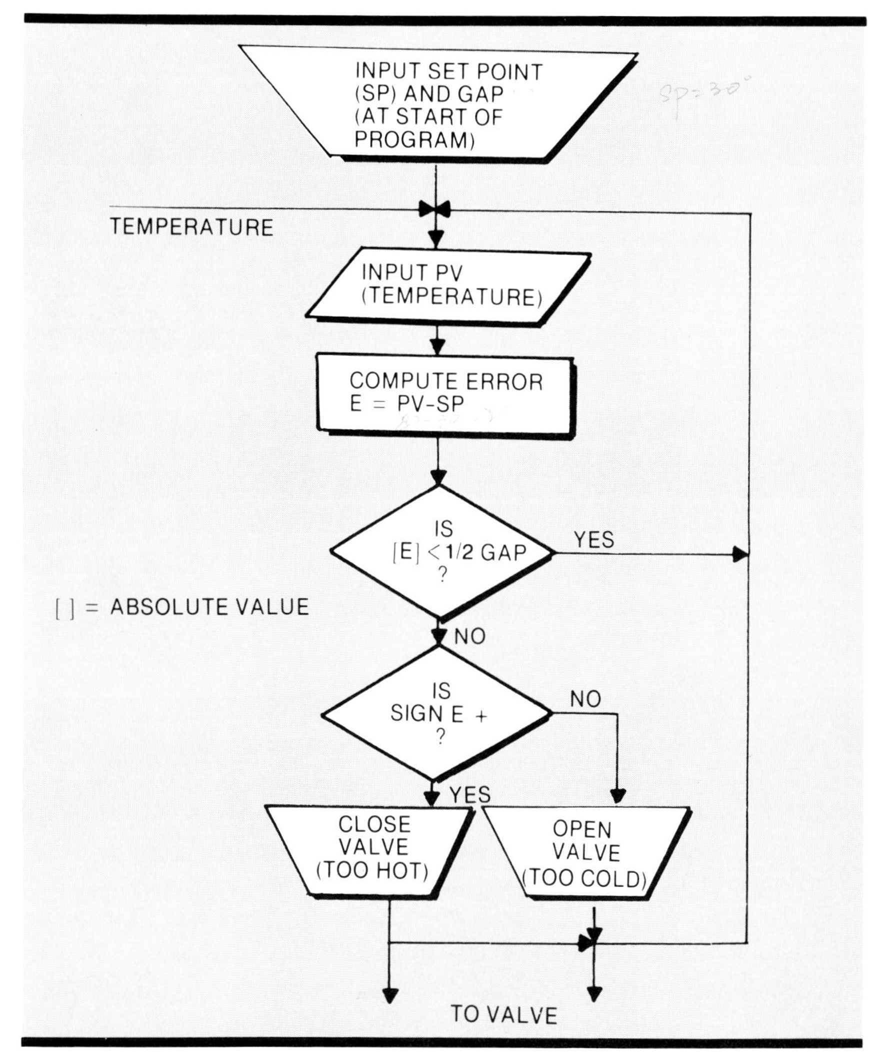

Fig. 7-2. Program Flow for On-Off Control

Example 7-1

The block diagram for an on-off control program (Fig. 7-1) is pictured in Fig. 7-2. The blocks represent operations, human or computer, expressed in mathematical or logical terms or in ordinary language. The lines represent flow of information or control.

The first block represents a manual input (or a prior decision) that establishes the value of the setpoint (SP) and GAP. (The trapezoid shape indicates input-output.)

The current temperature (PV) is measured and its numerical

value input to the second block. In the third block, PV is compared with SP, and the signed value of error, E = PV − SP, is calculated.

The fourth block is diamond-shaped, representing a decision. If the absolute (unsigned) value of the error, which we represent by [E], is less than half of GAP, no action is taken and control of the program returns to the "measure temperature" step. But if the error value puts PV beyond the GAP boundaries, the program flows to the next block.

The decision then must be made whether to open or close the valve. If the program has gotten this far, it already has been established that PV is beyond GAP so that only the direction of the error need be known to determine what action to take. The second decision block examines the sign of E. If E is positive, PV is too high and the valve must be closed. If the test for positive sign fails, then the valve must be opened, as PV is too low.

In either case the control is returned to the "input temperature" block and the cycle is repeated indefinitely.

The details portrayed in the flow diagram can be put in the form of an *algorithm*. An algorithm (named for the Arab mathematician al Khuwarizmi, A.D. 825) is a precise recipe or step-by-step procedure for solving a problem. An algorithm must meet the following criteria:

1. It is a finite and deterministic list of instructions to solve the problem.

2. It will work for all problems of the same class, regardless of the input.

3. The algorithm may be iterative, but the number of iterations is not known in advance of defining the inputs.

The algorithm for the on-off control problem is as follows.

Example 7-2

Step
1 Set GAP and SP to desired values

2 Input PV, calculate E= PV− SP
3 If (E) < GAP/2 go to step 2
4 If sign E is +, close valve, otherwise open valve
5 Go to step 2

An algorithm for direct computer implementation must use the elementary instruction of the computer. Note that we did not say how to convert temperature (PV) into a number; that is not within the scope of this unit.

A strictly mathematical procedure is given in the next example, how to compute the square root of a number N. It is based on a formula attributed to Newton.

$$X' = (X + N/X)/2 \tag{7-1}$$

where X' approaches the square root of N as the procedure is repeated with X' substituted for X.

Example 7-3

1. Start with an appropriate value of X such as N
2. Compute X'= (X+N/X)/2
3. If abs[X'− X] <e, X'= square root of N. Stop.
   (Note: e is any arbitrary small value.)
4. Let X= X', return to Step 2.

Example: solve for N =2, e =0.001

X=2:	X'=(2+1)/2=1.5	e=0.5
X=1.5:	X'=(1.5+2/1.5)/2=1.4167	e=0.0833
X=1.4167:	X'=(1.4167+2/1.4167)/2=1.4142	e=0.0024
X=1.4142:	X'=(1.4142+2/1.4142)/2=1.4142	e=0.0000

7-2. High- and Low-Level Languages

We return to the basic premise that the digital computer only understands the meaning of strings of binary digits, which it converts to a series of operations with buses, registers, and hardware like adders and shifters. Furthermore, these strings are only recognized as instructions when the computer is "told" that they are; otherwise they might be data: binary or decimal numbers or ASC symbols.

Binary code, or the hexadecimal equivalent, is thus the lowest level or *machine code*. It is possible for experienced programmers to read machine code but it is hardly convenient or obvious. Would you believe that 1000 1101 1110 0000 0011 means "store contents of accumulator at (decimal) address 1000?" How about "8D E8 03?" This may be made a little easier if you are told that the first byte is an instruction code (op code) which is always followed by 2 bytes of address. If that is still incomprehensible, it may help to know that the first byte of the address is the *low byte* in this instruction set, not the high-order numbers as we would ordinarily write them. The address 1000 or $03E8 is thus written in continuous memory locations as E8 03.

In early computer days, the front panel of every machine had a row of flashing LED or neon lights for operating and trouble-shooting use; these represented the binary contents of a register. (This kind of display is still beloved of Star-Trek and S-F movie producers.) Today, only the more expensive minicomputers and larger machines boast such displays, these being mostly in hexadecimal. The microcomputer has done away with these, both to save money and because the almost universal use of high-level languages makes it a rare event that machine code must be read. (However, in most microcomputers, the basic operating program, called a *monitor*, permits the contents of memory and registers to be displayed on the CRT screen, and to be altered as well.)

Earlier it was mentioned that the computer itself can be used to translate between machine code and something more comprehensible to the human user. For example, the instruction "8D" for the 6502 processor can be called STA, an easily recalled shorthand for "STore contents of Accumulator in memory." This kind of abbreviation is called a *mnemonic*, a code to aid memory.

An *assembly language* is a set of mnemonic codes that represents machine instructions. For every MPU, there is an assembly mnemonic for each type of instruction. However, instructions can be executed in one of several *addressing modes*; these will be explained further in this unit. Thus, for each mnemonic and applicable addressing mode there will be a unique op code. Programs written in assembly language must be translated or "assembled" into op codes before they will run on the machines. In order to determine which op code is applicable during assembly, some means of supplementing the mnemonics, such as punctuation or symbols, must be used to distinguish the addressing modes.

Assembly code can be translated into machine code by hand, a dreary, rote task of one-to-one translation. This is a good method to use for learning, but impossibly costly and slow for professionals writing long programs. Obviously, it is a perfect task for a computer! The computer program written to perform this translation is, not suprisingly, called an *assembler*. If we have an assembler program to use, we can write our application programs in assembly language mnemonics and turn them over to the computer for translation. The computer may be the *target machine*, the same on for which we are writing the program, or it may be a different one; in the latter case it is called a *cross-assembler*. Since the cross-assembler may be written for a larger and more powerful computer, it may support many programming conveniences such as good editor programs and disk-operating systems.

The assembly language program we write is called the *source program*. After translation into the 1s and 0s of machine code, the result is called the *object program*. This program can be assembled directly into RAM (if we are using the target machine for assembly) or it may be recorded on tape or disk and transported to the target machine.

Assembly language mnemonics are not standarized for all microprocessors. On the contrary, each manufacturer reinvents the wheel with his own variations. In fact, you can make up your own mnemonics if you insist. But, of course, it you want to use an assembler, you must use the code for which it was written, as well as the rules of spacing, punctuation, and symbols for addressing mode. In addition, each assembler has instructions for its own use, relating to the task of assembly rather than the MPU instructions: these are called *pseudo codes*. For example, the code .BA NNNN may mean "begin assembly at memory location NNNN."

To make things a little simpler, the examples in this and the next unit will be confined to one assembler and one microprocessor. The MPU is the 6502 (or any of the 6500 family) and the assembler is RAE (Resident Assembler/Editor), which can be used with the 6502 target machine, and was developed in ROM form by one of the major suppliers of the 6502, Synertek Systems Corp. The MPU mnemonics were developed by MOS Technology, Inc., the other major 6500 family producer. Experience with 6500 instruction sets and assemblers can be a step in learning the use of other instruction sets and assemblers for other MPUs, such as the 6800, 8080, and Z80, which are in general more complicated than the 6500.

The development of assembly languages and assemblers was a great step forward for the programmer, but many better things were in store. Assembly is not considered "high level" because of the one-to-one relationship to machine code, which is too detailed to support human concepts of mathematical and text manipulation. A statement like:

LET Y = X+3

can readily be understood by almost any grade-school graduate although it is written in perfectly valid BASIC, a widely used high-level programming language. (BASIC = Beginners' All-purpose Symbolic Instruction Code.) Each BASIC statement represents a whole series of op codes. The significance of high-level languages is that the programmer does not have to worry about machine operations but can concentrate more freely on his program application and logic. For this reason, they are often called *application or problem-oriented languages.*

There are advantages and disadvantages to high-level programming. The pros are speed and cost. It is estimated that writing a program in assembly takes ten times as much programmer effort as to write it in some high-level language. Also, many more people can use BASIC than assembly because it is easier to learn. In theory, high-level languages are *portable*, that is, they can run in another computer equipped to accept the language. Unfortunately, there are many different dialects of BASIC, for instance, and minor changes to the code may be necessary. High-level languages can usually be read and understood more easily by people who did not write them (or the people who did so, a year later) and so are more readily *maintained* and upgraded.

But there are disadvantages to using high-level languages, also. First, they require another step, a most complex one, in the inevitable translation to machine code. Two possible methods are used. One, the *interpreter*, is a program that must be resident in the computer memory, along with the source program, that will take each statement or line of code in a language like BASIC and convert it to machine code. Since the interpreter operates on a line-by-line basis, it is most convenient during program development, since each element of the program can be interpreted and debugged as we go along.

The alternative method is a *compiler*. This translates the entire source program to object form at one time, perhaps requiring several steps or "passes." Unlike the interpreter, which translates the source as it is being run and so must always be in residence, the compiler does its work before the program is run.

The disadvantage of the interpreter is the memory space it occupies. This may be 4K to 8K bytes for BASIC. FORTRAN (FOrmula TRANslation), a high-level language not unlike BASIC but more mathematically oriented, uses a compiler that may require as much as 32K bytes of memory. But since the object program, once compiled, no longer needs the compiler program, the memory space it occupied can be freed or the compiled program run on a machine with less memory.

High-level languages are problem oriented, so have their individual characteristics and strengths or weaknesses.

BASIC is a good all-around language and is easy to learn, but it is limited in mathematical and text-processing capability. It is often slow and needs a resident interpreter. (Newer BASICs can be compiled also.)

FORTRAN is better for mathematical work but is clumsy for text and needs a large compiler. Compiled code usually runs faster than interpreted and is more efficient, sometimes optimized by the compiler. RATFOR (RATional FORtran) is a highly structured form of FORTRAN (see structure discussion, below).

FORTH is an extremely versatile and efficient language that is especially good for control problems using small computers, but it is difficult to learn and read (almost like assembly). It requires a large compiler, but one that is *incremental,* that is, each statement is compiled as written and put into a "dictionary." Each definition refers to prior definitions, so that the process of interpretation means "threading" back to the source, therefore FORTH is called a *threaded code*. The final program is usually one definition, encompassing all prior ones.

It usually can be run with a much smaller dictionary than that used for the original compilation, so that it acts like an interpreter.

Pascal, named for the French mathematician, is an excellent logical language and is highly *structured*. This means it enforces rigorous discipline on the programmer and requires programs to be written in a formal sequence of operations (heading, define variable types, define subroutines, main program). For these reasons, it is supposed to be easy to teach and learn and prevents logical errors.

Pascal requires a large (20K byte) compiler and, for microprocessors, a small resident interpreter for an intermediate computer-produced source (called P-code).

COBOL (COmmon Business Oriented Language) is designed for such tasks as file processing and updating, billing and payroll. It is poor for mathematical problems. It requires a large compiler, so is not too practical for a small computer.

ADA is a relatively new language, named for an English countess who was the first programmer for Babbage, an early computer inventor. ADA is much like Pascal but is more complex and so is oriented toward 16-bit and larger processors. It is optimized for military and large-control problems.

PL/1 languages (including PL/m, PL/80, MPL, and others) are based on an earlier Pascal predecessor, an IBM development called PL/1 (for Programming Language 1). The various offshoots of PL/1 have been adapted to the 8080 and Z80 microprocessors and include a variety of propriety dialects.

In choosing a language, the degree of structural discipline it imposes on the user should be considered. In a structured language, only certain kinds of programming constructs are allowed. This makes code easier to read and maintain and more reliable. Typing variables, that is, limiting certain variables to specific types such as letters of the alphabet or integers 1 to 10 or days of the week also promotes reliability. If wrong data is inserted, it usually can be spotted. Languages like ADA and Pascal are the most structured of those in use today: BASIC and FORTRAN are less so, and FORTH or assembly languages least of all.

All high-level languages, then, have their strong and weak points which reflect the class of problems for which they were designed. With the possible exception of FORTH, the code produced by all high-level languages is considered less efficient in terms of memory used and operating time than assembly language. This is because the object code-producing mechanisms of compilers and interpreters are less able than people to take advantage of clever shortcuts and patterns.

Another significant point to be taken when considering high-level languages *vs.* assembly for control and instrument works is the comparative ability to handle bits within individual words or bytes. Machine language is far better suited to this kind of manipulation than high-level types. Bits are important in control work where the status of switches and on-off controls are monitored and altered by the computer program. The control of computer input and output peripherals, such as printers, through interface chips, also requires bit setting and monitoring. To do this with BASIC or FORTRAN is possible, but has been compared to "eating soup with chopsticks."

7-3. Instruction Sets and Addressing Modes

Assembly language and assemblers work with mnemonics that represent op codes, the latter being directly usable by the MPU when expressed in binary form. Assemblers also accept addresses and data in hexadecimal and usually decimal, binary, and even ASCII form. Thus, every statement in assembly code can be translated on a one-to-one basis to machine or binary code. The process of translation (assembly) can be done by hand or (as we will see) by the assembler program and the computer itself.

Each MPU has its own unique list of instructions and op codes, known collectively as the *instruction set.* We will confine ourselves at this time to a few instructions of the 6502 MPU, a microprocessor used on many personal computers that might be available to you. If at all possible, you should obtain access to one of these computers, such as the Apple, the Commodore PET, Atari, or OSI, preferably one that is equipped with an assembler, and use it to practice programming. Assembly programming cannot be learned from books; it must be practiced.

Less expensive and even more suitable for learning machine programming are the single-board computers such as the KIM (Commodore), the SYM-1 (Synertek), or the AIM (AMD).

These are boards equipped with a 6502 MPU, an expandable minimum of RAM, a hexadecimal or standard keyboard, and a one-line display. They can be programmed directly in hex machine language without an assembler. For checking the programs in this text, the writer has used a SYM-1 equipped with a keyboard terminal (Synertek KTM-2), a surplus CRT monitor, and an assembler-editor in ROM (RAE-1). With this setup one can write and edit programs in assembly language and assemble them starting anywhere in memory. The assembler also notifies the user of certain syntax errors, prints listing of the program in assembly and machine code, and produces a list of *labels* and their addresses (labels will be explained later). Programs can be recorded and stored on tape or disk. Thus, this fairly elementary equipment can be called a *development system* since it is adequate for program development.

At this point we will discuss the anatomy of a 6502 instruction, similar to one used in other MPUs, and explain the addressing modes used. In the next unit we will follow through with a small program using this instruction and demonstrate the effect it has on the registers and timing unit.

(Note that the mnemonics used are unique to each manufacturer and are generally copyright subject. When we use 6502 mnemonics it should be understood that they are "copyright by MOS Technology, Inc., 1975, all rights reserved.")

LDA — Load Accumulator with memory

You will recall that the 6502 accumulator is a general-purpose register that interfaces the data bus directly, as does the memory, Fig. 6-2. It can, therefore, be used to transfer data to memory or the reverse, or serve as intermediate storage between two memory locations. The accumulator also interfaces the ALU and so can serve both as an ALU input and a temporary store for the results of its calculations, such as addition.

The mnemonic LDA directs the transfer of the contents of a memory address to the accumulator. Note that this is only one small step in any useful program. To process these data in the

ALU or to transfer to another address requires at least one more instruction. This step-by-step procedure is typical of assembly and machine programming, and, indeed, of the operation of the MPU itself.

But merely stating the LDA instruction is not enough. Where will we find the data to load into the accumulator? This introduces the problem of *addressing mode.*

Reference to the *6502 Programming Manual* (Ref. 1) gives the following information (Table 7-1):

Addressing Mode	Assembly Language	Op Code	No. Bytes	No. Cycles
Immediate	LDA #oper	A9	2	2
Zero page	LDA *oper	A5	2	3
Zero page, X	LDA *oper, X	B5	2	4
Absolute	LDA oper	AD	3	4
Absolute, X	LDA oper, X	BD	3	4(a)
Absolute, Y	LDA oper, Y	B9	3	4(a)
(Indirect, X)	LDA (oper, X)	A1	2	6
(Indirect), Y	LDA (oper), Y	B1	2	5(a)

Notes: Oper signifies operand, the value specified in the field following the mnemonic; (a) Add 1 if page boundary is crossed.

TABLE 7-1. LDA Instruction (Ref. 1)

It is clear from this table that the assembler needs to know the addressing mode before assigning an op code to LDA since there are eight choices. These are, in fact, all of the 6502 addressing modes, so that LDA is termed a "Group One" instruction; other groups use fewer modes. But what do they signify? Why, for example, don't we just specify the 16-bit address in the next 2 bytes following the op code and let it go at that?

In fact, this is one choice, called the *absolute mode,* and has the op code AD. Examining the table, however, reveals that this is not necessarily the best choice and would severely constrain the programmer. The reason for this statement is given in the last two columns. We use machine-level programming because it is more efficient, meaning that *memory and time* are more efficiently used. The absolute mode takes *3 bytes*, one for the op code and 2 bytes for address. Further, it requires *four cycles* to complete the instruction. The table shows that this is not the best we can expect. Under some circumstances, it is possible to get the address with only 1 byte (two total for the instruction), and the number of cycles can be as few as two or as many as six.

Immediate addressing is both the most efficient mode and the simplest. It instructs the processor to load exactly the next byte following the op code and to consider it as *data*. In other words, the address of the data to be fetched is always the address immediately following the op code byte itself. In order to specify the immediate mode, the assembler requires the numeral sign # to precede the operand. For example, LDA #0 will load the byte of data 00 into the accumulator, while LDA #$1A will load the hex value $A1. If we were to observe the contents of the memory locations storing these two instructions, we would see the following code in these two cases:

```
A9 00
A9 1A
```

(Note that the assembler requires the dollar sign $ in order to indicate hex data. Binary data is preceded by the percent sign (%). The default, if no sign is used, is decimal.)

Immediate addressing is the simplest way to manipulate a *constant*, which is defined as a "value known to the programmer." This mode requires only a minimum time to execute since the first cycle loads the op code and the second cycle fetches the constant while the op code is being interpreted. Absolute addressing, on the other hand, requires four cycles. On the first, the instruction is fetched. The second and third fetch the high and low bytes of the address from contiguous memory following the op code. On the fourth cycle, both bytes are put on the address bus and the data is returned from memory and loaded into the accumulator.

The next most efficient mode after immediate is *zero page addressing*. Zero page is the block of memory (page) from $0000 to $00FF and has a very special significance in the 6502. This is because any data placed in this page can be retrieved with this special addressing mode with a savings in space and time amounting to 1 byte and one clock cycle (versus the absolute addressing mode). Use of zero page can, therefore, result in significant savings if the program is long or repetitive. This savings is achieved by the 6502 because the op code A5 (which is called by the assembler when it sees the asterisk * symbol before the operand) automatically places a 00 in the high-byte address register, so that only the low byte need be fetched. Note that the difference between zero-page addressing and immediate is that the operand in the first case is an address, and the contents of that address (which end up in the accumulator) is a *variable*. The operand in immediate mode, however, is a *constant* which is placed in the accumulator, and will not change during the program. Thus, while both are 2-byte instructions, they have very different uses.

Because there are only 256 addresses in a page, the memory space in zero page is limited and valuable, hence it should be reserved by the programmer only for the most important data—that which must be frequently fetched. One such valuable use for zero page is to store *pointers*. A pointer is merely 2 contiguous bytes containing the address of another location, which may be anywhere in memory, and that holds the data ultimately desired. If a pointer is frequently used, as for a table of data or the address of a routine, it can be accessed more quickly in zero page than anywhere else. We could not use immediate addressing because we need two bytes for an address. Furthermore, it is possible to change the data which is being pointed to without changing the part of the program that contains the pointer, thus making it easier to change or debug a program.

7-4. Indexed and Relative Addressing

Absolute *indexed addressing* is shown in Table 7-1 as "Absolute, X" and "Absolute, Y." Indexing introduces a new concept, that of the *computed address*. The address to which LDA goes for data is computed by adding to the absolute (or zero page) value, called the *base address*, the value of another register, X or Y, which is used as a counter. With this idea, it is possible to address an entire block of contiguous addresses merely by incrementing the X or Y register. (An example of this technique will be given further on.)

The base address in indexed addressing often is expressed in terms of a *symbolic expression* or just symbol, rather than a fixed address. Thus, the expression TABLE instead of $1000 can be used for a list of data that begins at address $1000. Most assemblers permit this. Of course, the symbol TABLE must be defined to the assembler so that it can be converted to a real binary number in machine code. This is accomplished by an instruction called a *pseudo op* which addresses the assembler program, not the microprocessor. For example, some assemblers use the *equate pseudo op*:

TABLE = $1000
or TABLE EQU $1000

In RAE the correct expression is:

TABLE .DE $1000

if the location is outside (external) to the program addresses, or

TABLE .DI $1000

if it is part of the program address space.

With this definition, we can refer to the symbol TABLE anywhere in the program or even use an expression such as TABLE+1 or TABLE+2. Symbols add a powerful capability to the assembly language for two reasons; first, the address of the symbol can be changed easily wherever it is used in the program merely by redefining the pseudo op. This makes it easy to *relocate* (move) the program to another part of memory. Second, the symbol assigned can be a name that is meaningful in ordinary language or to the programmer. This tends to make the program more "friendly" and gives assembly some of the advantage of a high-level language.

The assembler keeps a list of these symbols and also of *labels*, which are another kind of symbol used to name the address of a program instruction. These lists are printed out by the assembler on demand.

Indirect addressing is a form offered by the LDA instruction that is used mostly by relatively sophisticated programmers, so we will only mention it in passing. The basic idea of an indirect address is that the effective address of the desired data is actually in zero page, and the instruction actually contains only the low byte of that zero-page address following the op code. Thus, the MPU's address register is first referred to a location in zero page, which it uses as a pointer. This location contains the low byte of the real address; the next zero-page location contains the high byte. Both are then put into the address register and locate the true location of the data. The major use of indirect in the 6502 is picking up data from tables or lists, thus the instructions have also an "indexing" feature which allows one to add the X or Y register value to the zero-base address. This is the *indexed indirect* form. The other type of indirect is *indirect indexed* addressing. In this case, the indirect base address is first located as in the previous method, and only then does the indexing feature take effect. Due to "pipelining" of microinstructions in the 6502, one cycle of clock time is saved by these procedures.

7-5. Jump and Branch Instructions

Only one 6502 instruction has a "pure" or absolute indirect addressing mode; that is JMP, the *jump* instruction. JMP changes the contents of the address register according to the two bytes following the JMP op code (address LO and address HI). Therefore it is 3-byte instruction. If the op code is 4C, the jump is absolute, that is, the address register directly adopts the new address bytes. If the op code is 6C for JMP (Oper) the next two bytes are a pointer containing the actual jump address. For example, JMP ($1000) causes the processor to fetch the contents of address $1000. Suppose $1000 contains $40 and the contents of $1000+1 are 3A, then the jump address is $3A40.

There is one additional form of addressing that is not found in the LDA table or, indeed, any 6502 instruction other than branch operations. This is *relative addressing*, where the address register is merely advanced or decreased by an amount specified in the instruction (within the limits +128 to −127). The byte following the branch code is treated as a signed hexadecimal number, consequently it cannot exceed the limits given above. The number represents the movement of the address register, ahead or behind, *relative* to the current position, whatever that may be. The upper limit on branching constitutes a definite limitation of the 6502 processor which is not shared by all other MPUs. If we wish to branch a distance greater than the limits, we must arrange to branch to an intermediate address within the limitation; that address can contain an absolute JMP instruction which is not limited.

The most important characteristic of the branch instructions is that they are *conditional*, that is, whether or not the branch is taken depends on the status of certain flags in the status register.

Some of the branch instructions available in the 6502 are listed in Table 7-2.

Mnemonic	Op Code	Results
BMI	30	Branch on result minus (N bit set)
BPL	10	Branch on result minus (N bit zero)
BCC	90	Branch on Carry clear
BCS	B0	Branch on Carry set
BEQ	F0	Branch on result zero (Z bit set)
BNE	D0	Branch on result not zero
BVS	70	Branch on Overflow (V bit set)
BVC	50	Branch on Overflow clear (V bit 0)

TABLE 7-2. Branch Instructions

The branch takes place depending on the result of the previous operation. For example, if two numbers are subtracted and they are equal, the result is zero, the Z bit is set in the status register, and if the next instruction is BEQ NN, the program will jump NN locations from the current program counter. Whether or not the branch is taken, none of the branch instructions affects any flags or registers except the program register. If the branch is not taken, the instruction cost two cycles of clock time. If the branch is taken, one cycle is added, or two if the branch crosses a page boundary. (It should be mentioned that the LDA instruction does affect the zero flag, setting it if the accumulator is zero, and also the negative flag if bit 7 of the accumulator after LDA is 1. No other flags or registers are affected.)

7-6. Implied Addressing and Summary

An implied address stipulates an operation that is internal to the processor, that is, no address need be stated. For example, the instruction may clear or set a bit in a register, such as CLC (clear Carry bit), CLD (clear decimal flag), or SEC and SED to set the carry and decimal flags. Since the instruction itself "implies" the address, i.e., the register, none is required. Therefore, these instructions require only a single byte. Another type of implied address instruction increments or decrements the contents of a register, such as INX or INY to increment the X and Y registers respectively, or DEX and DEY to decrement them. Still another type for which no address need be stated are those that transfer bytes between registers, for example, TAX and TAY transfer the accumulator contents to X or Y and TYA, TXA transfer them back. The stack pointer can be transferred to the X register and restored using the instructions TSX and TXS.

Since each of these instructions require only 1 byte they are very economical.

Many complex operations involving saving (pushing) register on the stack and popping them back require more than two cycles. These include BRK (break), also called a forced or software interrupt, PHA and PLA which push and pop the accumulator from the stack, and others.

The *accumulator mode* is much like the implied. It refers to instructions that shift and rotate bits in the accumulator (LSR or logical shift right, for example). These instructions may operate either on the accumulator or on memory registers. Only 1 byte and two clock cycles are needed if the accumulator is implied, otherwise additional bytes are needed for the memory address, as well as more clock cycles.

Summarizing, there eleven modes of addressing used by the 6502 MPU. Many of these are not used except by advanced programmers. The beginning assembly language programmer is advised to use only those instructions and addressing modes which he or she understands well and only gradually expand one's "vocabulary."

The eleven modes, with examples, are summarized in the following table.

Mode	Example	Op Code	Cycles
Absolute	LDA	AD	4
Absolute, indexed (Y)	LDY	BC	4*
Accumulator	LSR	4A	2
Immediate	LDX	A2	2
Implied	INX	EB	2
Indexed, indirect (X)	LDA	A1	6
Indirect	JMP	6C	3
Indirect, indexed (Y)	LDA	B1	4*
Relative	BEQ	F0	2**
Zero page	DEC(a)	C6	5
Zero page, indexed (X)	AND(b)	35	4

* Add 1 if page boundary crossed.
** Add 1 if page boundary and 1 if branch taken.
a Decrement memory.
b AND to the accumulator contents.

TABLE 7-3. Addressing Mode Summary

One more instruction is required before we can follow the results of the simple program presented in the next unit.

The STA instruction (Store Accumulator in memory) is the converse of LDA and has all of its addressing modes except one, the immediate. This instruction is listed in Appendix B.

Exercises

7-1. *Using the algorithm of Example 702, calculate the square root of 10 (decimal) to a precision of e=0.001. How many iterations are required?*

7-2. *A cake recipe specifies the amount of each ingredient, the mixing method, the baking time, and the temperature. Is this an algorithm? Is it a complete program?*

7-3. *Draw a flow diagram for the square root algorithm. Use your own words in the boxes, but the standard shapes shown in Fig. 7-2.*

7-4. *What programming language or languages would you recommend in the following cases:*

   a) *A small microwave oven control with a simple control program (less than 500 bytes), to be produced in large quantity.*
   b) *A long and complex process-control optimizing problem which involves much mathematical calculation and must be reliable.*
   c) *A specialized, experimental controller to be used in your plant. The program is not large but must be capable of being modified and updated frequently.*
   d) *A small program must be prepared for a personal computer. You don't know any programming language.*

7-5. *Will the increasing use of 16-bit microcomputers affect the choice of application-programming languages? How? Why?*

7-6. *Give the 6502 instructions for Store Accumulator in Memory location 1500 (decimal) in assembly, hex, and binary codes. (Note: try to do this without looking back at the example in this section.)*

7-7. *The names of BASIC variables are PV (for process variable), SP (for setpoint), and ER (for error). Guess what line of code you would write in BASIC to define error. How many lines of assembly code do you think that would take?*

7-8. *Give the reason for the convention of storing an address as "lo byte, hi byte" instead of the reverse, as you would expect from our ordinary method of writing most significant numbers first. (Hint: consider zero-page addressing.)*

7-9. *Write the 6502 assembly code to put into the accumulator a "mask" to test bit 6 by ANDing. Use:*
a) binary notation;
b) hexadecimal.
What is actually stored in the program memory?

7-10. *Write the assembly code and hex code to load into the accumulator*
a) the constant 9;
b) the contents of $00FE;
c) the contents of $40A5.
What is the penalty for using LDA $FE00 as a solution to (b)?

7-11. *Explain why STA does not have an immediate mode.*

References

[1]*SY6500/MCS6500 Microprocessor Programming Manual,* Publication 6500-50. Santa Clara, California: Synertek, August 1976.

Unit 8: Assembly Language Programming

Unit 8

Assembly Language Programming

In this unit we will consider assembly language programming, starting with an elementary program segment and leading up to a fairly complex working program.

Learning Objectives—When you have completed this unit you should:

A. Understand the details of an elementary program segment and the response of registers to instructions.

B. Know how to use an assembly language and some instructions of an MPU instruction set.

C. Recognize the use of some basic programming techniques, including loops, subroutines, and labels.

8-1. Writing Assembly Programs

Throughout this unit we will continue to use the 6502 mnemonics and the RAE assembler code as examples. Addresses are in hex code.

Programs written in assembly language should strive to be simple, clear, and well structured. At the beginning, you should not attempt to write the fastest and most efficient program, but the one most easily understood. Refinements in speed and memory use should come later. To emphasize clarity, you should use comments and labels. Comments in many assemblers follow a semicolon in the last field of the statement and should be in plain language or readily recognized abbreviations. It is considered bad form to explain the meaning of instructions in comments (that's what the mnemonics are for). Instead, say what is happening to the data. (We will occasionally break this rule here in the interest of clarity.) Labels, as explained earlier, make a program relocatable in memory as well as more easily understood. Comments are ignored by the assembler and are not loaded into memory with the object program. Instead, they are converted to absolute addresses, which will depend on where the program is assembled.

When writing programs for an assembler we must follow its rules of *format*. These may differ between assemblers. For RAE, we have five fields, each, except label, separated by a blank space. These are:

Line No.	(Label if any)	Mnemonic or Pseudo Op	Operand	;Comment

Each line (called a *program statement*) must have a line number for purposes of ordering and editing. The line numbers can be arbitrary but must be in ascending sequence. Lines can be replaced by retyping with the same line number or inserted between other lines by using an intermediate line number. Consequently, it is wise to number lines in multiples of 5 or 10 so as to leave room for later insertions. Program statements can be written in any order and will be sorted correctly by line number by the assembler.

After writing and editing a program, it is assembled at a specific location in memory specified by an assembler pseudo op. The program is then converted to object (machine) language and can be *listed* (printed out) with the machine codes and addresses in hex. The output listing format is:

Address	Machine_Code	Line_No.	(Label)	Mnemonic	Operand	;Comment

Note that the assembled code address is the location of the machine code, not the assembly (source) code. In RAE the source code starts on page two, since zero page is saved for important references and one page is the stack.

8-2. Loading Data into Memory

The objective of this program segment is simply to put data into a memory location. In addition, we will demonstrate what happens "inside the MPU" when the program is loaded and run. The input datum is the hex constant $A0. It is first loaded into the accumulator with the instruction LDA in the immediate addressing mode. Then the accumulator contents are transferred to the selected memory location using the instruction STA in the absolute mode. The assembly code typed into the computer would appear thusly:

Example 8-1

```
090              .OS            ;PSEUDO OP
100START         LDA #$AO       ;INPUT CONSTANT
110              STA $1000      ;STORE IN MEMORY
120              BRK            ;RETURN TO MONITOR
130              .EN            ;PSEUDO OP (END)
```

It's assumed that the Break code BRK (Hex 00) returns control of the processor to the monitor program (the basic operating system of the computer). The .EN code on line 140 is not an MPU instruction but is an assembler program command. It indicates to the assembler the end of the program. Similarily, the .OS on line 90 is another pseudo op, meaning "object store." Without .OS the assembler would go through the motions of assembling, including looking for errors, but would not actually store the object program in memory. The default address for object storage (in the RAE) is $0200. If another starting address is wanted, say $0500, we would add the following line of code:

```
080    .BA  $500     ;BEGIN ASSEMBLY AT $500
```

If we now type the assembler command ASSEMBLE LIST we will be rewarded with the RAE listing:

```
0200- A9 A0      100START  LDA  #$A0     ;INPUT CONSTANT
0202- 8D 00 10   110       STA  $1000    ;STORE IN MEMORY
0205- 00         120       BRK           ;RETURN TO MONITOR


LABEL FILE [/ = EXTERNAL]

START= 0200
```

Recalling the listing format we see that the first four columns (to the left of the line number) are devoted to the object code beginning address and the hex code for each instruction. The label file has only one item, START, which was assigned the absolute address $0200 by the assembler (since we did not use the .BA pseudo op).

If each memory location starting at $0200 was examined, using the monitor program which has provisions for doing this, we would see the following:

```
0200    A9
0201    A0
0202    8D
0203    00
0204    10
0205    00
```

where the first column is the hex address and the second the contents of the memory cell at that address.

Another form of machine code display given by most monitors is the "Verify" or "Hex dump" form. A hex dump begins at the specified address and prints 8 or 16 bytes in sequence, followed by a checksum, viz:

```
0200 A9 A0 8D 00 10 00 00 00, E6
01E6
```

(Since the program is only 6 bytes long it is assumed to be padded out by additional BRK codes, although anything can be located there.) The checksum is the hex sum of all code digits, adding to $10E6, while the checksum for the line only is the last two digits of the line sum. The checksum allows the programmer to see quickly if he has made any errors in copying a line of code without having to check every byte.

The assembler program did not do anything you could not have done by hand. In this example it would be easy to write the object-code program directly from the source code of Example 8-1 and to enter it into memory using the utilities of the monitor. However, it is obvious that this would become very tedious for a program of many hundreds or thousands of bytes!

The program of Example 8-1 can be *run* by using the command RUN $200 if in the RAE editor program, or .G 200 if in the monitor. Since it is instructive and helpful to good programming to be able to visualize what is happening when the program runs, let us imagine we can visualize what is happening when the program runs, let us imagine we can visualize the sequential changes in all the registers for this example. In actuality, since each clock cycle is only one microsecond (or less), this can be observed only with special instrumentation.

Table 8-1 lists the changes in address and data buses with each clock cycle after RUN or .G. The "external operation" column tells what data is on the buses and the "internal operation" is what is going on in the MPU hardware.

Clock Cycle	External Operation	Address Bus	Data Bus	Internal Operation
1	Fetch op code from memory	0200	A9	Increment PC (Program counter) to 201
2	Fetch data from memory	0201	A0	Interpret A9 instruction Increment PC to 202
3	Fetch next op code from memory	0202	8D	Load data in accumulator Increment PC to 203
4	Fetch first address half from memory	0203	00	Interpret 8D instruction Increment PC to 204
5	Fetch second half address	0204	10	Hold address lo byte (ADL) Increment PC to 205
6	---	1000	A0	Put full address on PC Fetch data from accumulator
7	Fetch op code from memory	0205	00	Load data (accumulator) in memory location

TABLE 8-1. Program to Load Data into Memory

Figure 8-1 is a graphical representation of the memory and of the pertinent registers (called a *programming model*) as their contents change from cycle to cycle. At the left is pictured the memory contents. At the right are the registers and their contents.

The memory contents from $200 to $205 hold the object code. In the first clock cycle, code A9 is read from $200 and put into the instruction register. The program counter PC is automatically advanced to $201. The contents of the accumulator (A) and data latch (DL) are not known and don't matter at this point.

During the second cycle, the op code (A9) is deciphered, and since it is found to be a 2-byte instruction (immediate addressing mode) the second byte of the program is treated as data and is brought into the data latch. During the third cycle, the A9 instruction is completed when the datum A0 moves to

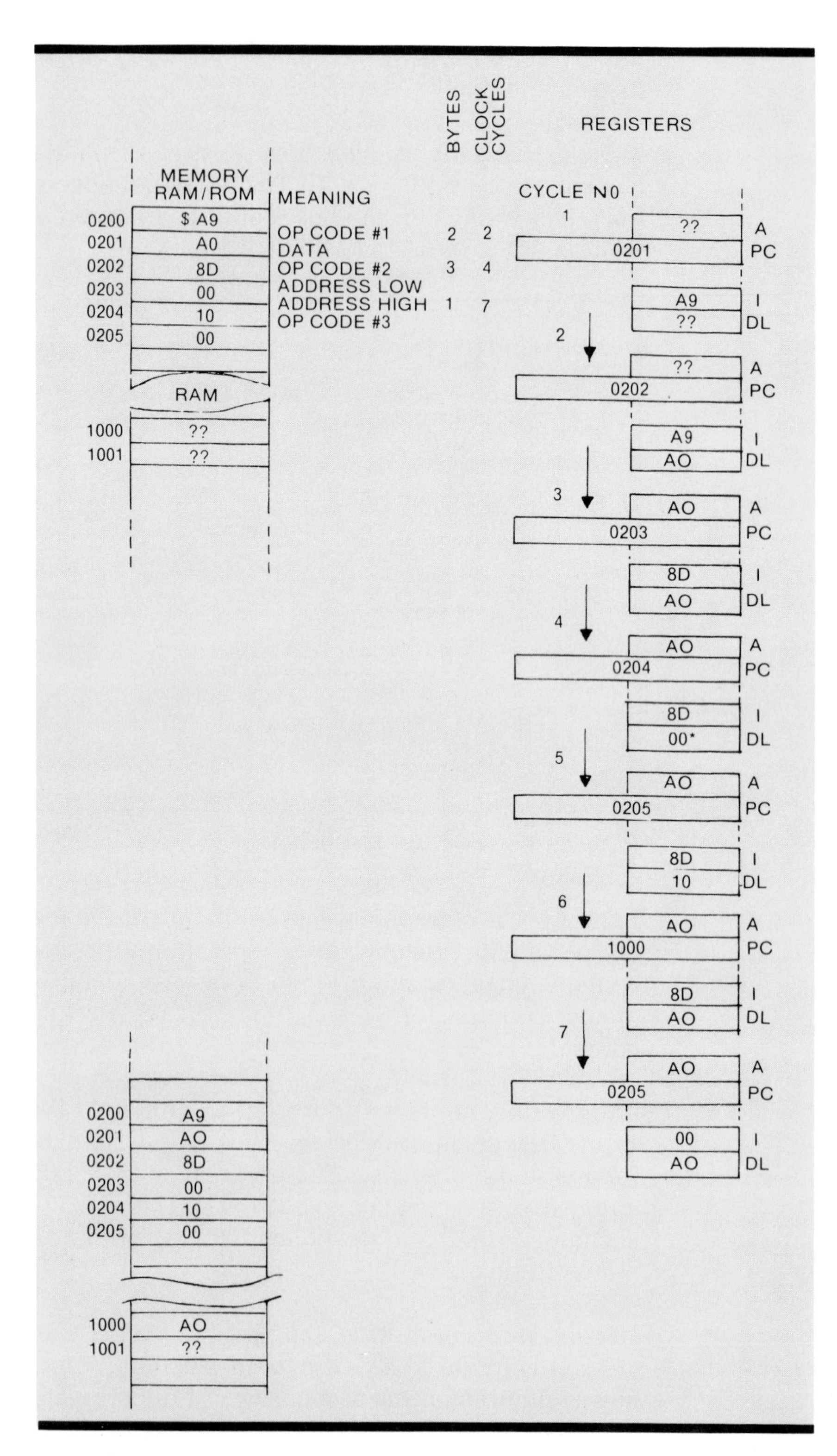

Fig. 8-1. Programming Model for Example 8-1

the accumulator from DL, and at the same time the next op code is fetched.

The op code 8D is a 3-byte instruction, hence first the low byte of the absolute address (00) is brought into the data latch and held until the next cycle, when the high byte (10) will be brought in (fifth cycle). In the sixth cycle, both bytes are put on the counter PC while the datum held in the accumulator is put out on the data bus. Note that the contents of the accumulator do not change. In the seventh cycle, the last op code is brought in and terminates the program.

8-3. Addition of 8-Bit and 16-Bit Numbers

For a second example, we examine a program to add two unsigned 8-bit numbers. The result must be less than 256 since we will make no provision for handling overflow from the 8-bit register. We will need two new instructions for this operation.

CLC Clear Carry Flag Bit
Addressing mode — implied
Op Code 18 1 byte 2 cycles
Status flags affected: C (=0)

and ADC Add Memory to Accumulator with Carry
Addressing modes — (see Appendix B)
Op Code (for zero page) 65 2 bytes 3 cycles
Status flags affected: N Z C V

We also will assume that the addend is to be found in zero page at location $0010, that it is to be added to the contents of $500 and stored at location $600.

Example 8-2
Unsigned 8-bit Addition.

```
                         SOURCE CODE
Clock
Cycles(a)
  2      100   CLC         ;CLEAR CARRY
  4      110   LDA $500    ;GET AUGEND FROM ABS.ADDR.
  3      120   ADC *$10    ;ADD ADDEND FROM ZERO PG.
  4      130   STA $600    ;STORE RESULT IN MEMORY
 ___
 13
```

a) Does not appear in listing—for information only.

Address (hex)	Contents Before Run	After Run
0010	15	15
0500	2A	2A
0600	??	3F

Object	Program
200	18
201	AD
202	00
203	05
204	65
205	10
205	8D
207	00
208	06

Example 8-3
16-Bit Addition.

The 8-bit registers of the 6502 microprocessor can perform double-precision arithmetic (16-bit) for 16-bit results. But additional time and program space must be allocated, since the high and low bytes must be added separately with attention to the carry bit from the lower bytes. The Add with Carry (ADC) instruction of the 6502 is suitable for this task.

Assume that we wish to add the two unsigned 16-bit numbers 1A20 and 03FF. The result should be 1E1F (decimal 7711). Six memory locations must be reserved for the 2 bytes of the first number (L1, H1), the second number (L2, H2), and the 16-bit results (L3, H3). Assume also that the zero-page location $0010 through $0015 are available for this purpose. These locations will hold the following:

Address	Number	Addition Before	After
0100	L1	20	20
0011	H1	1A	1A
0012	L2	FF	FF
0013	H2	03	03
0014	L3	?	1F
0015	H3	?	1E

The source program is:

```
100  LDA *$10      ;GET LOW ORDER BYTE FIRST NUM.
110  CLC           ;CLEAR CARRY
120  ADC *$12      ;ADD TO LO BYTE 2ND NUM.
130  STA *$14      ;STORE LO BYTE IN RESULT LO
140  LDA *$11      ;GET HI BYTE 1ST NUM.
150  ADC *$13      ;ADD 2ND HI BYTE WITH CARRY
160                ;FROM LINE 120
170  STA *$15      ;STORE HI BYTE IN RESULT HI
180  BRK
```

8-4. Loop Program

A *loop* is a repetitive performance of some program segment. Multiplication, for example, can be accomplished by repeated addition, for instance, 5 times 9 is the sum of 9 added 5 times.

The multiplier, the number of additions to be accomplished, can be put into an index register and the register contents decremented each time the addition operation is performed. When the value of the index register is reduced to zero, the cumulative sum of the additions is the desired product. A general picture of loop operations is shown in Fig. 8-2.

Similarly, a loop can be used as a delay timer, a very handy operation in many control problems. The 6502 has two index registers, X and Y, and commands to load and decrement them.

LDX Load Index X with memory contents

LDY Load Index Y with memory contents

DEX Decrement Index X by one

DEY Decrement Index Y by one

Both DEX and DEY affect the Z register. Therefore, the instruction BNE (Branch on result not zero) can be used to test the registers after each decrement to see if they have yet reached zero.

Assume that we wish a delay of approximately Y milliseconds. A program that will accomplish this is the following: (see Fig. 8-2.)

Example 8-4
Time Delay Program.

```
100DELAY     LDX #MILSC    ;COUNT FOR 1 MS DELAY
110DLYX      DEX           ;XINDEX = XINDEX-1
120          BNE DLYX      ;IS INDEX = 0?
130          DEY           ;YINDEX = YINDEX-1
140          BNE DELAY     ;COUNTED NO. MS YET?
150          BRK
```

The Y index is initially loaded with the number of milliseconds delay wanted, using the instruction:

```
LDY  nn  ;nn = NO. OF MILLISEC
```

The constant MILSC is the number of X loops needed for 1 ms delay. To determine its value, consider the clock delays of the program. For each iteration of the X index loop, the clock periods are:

DEX 2

BNE 3 (or 2 when branch not taken.)

Crossing page boundaries has not been considered in the above. Thus, there are approximately five clock cycles per iteration of the loop. At a standard clock rate of one MHz there are 1000 clock periods per ms. The number of iterations equal to MILSC is then 1000/5 or 200 (=$C8 hex). (Due to branches taken and the LDX instruction, the above is not exact; a closer value is 199. But there are other factors that may cause greater errors, such as temperature effects, hence $C8 is close enough.)

8-5. Shift Operations

The shift and rotate instructions have been discussed earlier. The ASL (Arithmetic Shift Left), for example, enables one to save the MSB of a byte by shifting it into the carry position. Since bit 7 is also the sign bit, this means of saving or testing the sign of an operation.

The following program segment gets data from location $10 and performs a left shift, storing the result in $11.

Example 8-5

	BEFORE		
ADDRESS	SHIFT		AFTER
0010	7F	7F	
0011	??	FE	
100	LDA *$10	;PUT DATA INTO ACCUMULATOR	
110	ASL A	;SHIFT ACCUMULATOR LEFT	
120	STA *$11	;STORE RESULT	

Note that the result placed in $0011 is twice the original value. Do you know why? If a left shift is equal to a multiplication by two, this enables us to develop an algorithm for multiplication by any binary number; more elegant and faster than repeated addition. The multiplier is shifted left and each bit going to the Carry is examined. If Carry = 1, the multiplicand is added to the product. Otherwise, nothing is added.

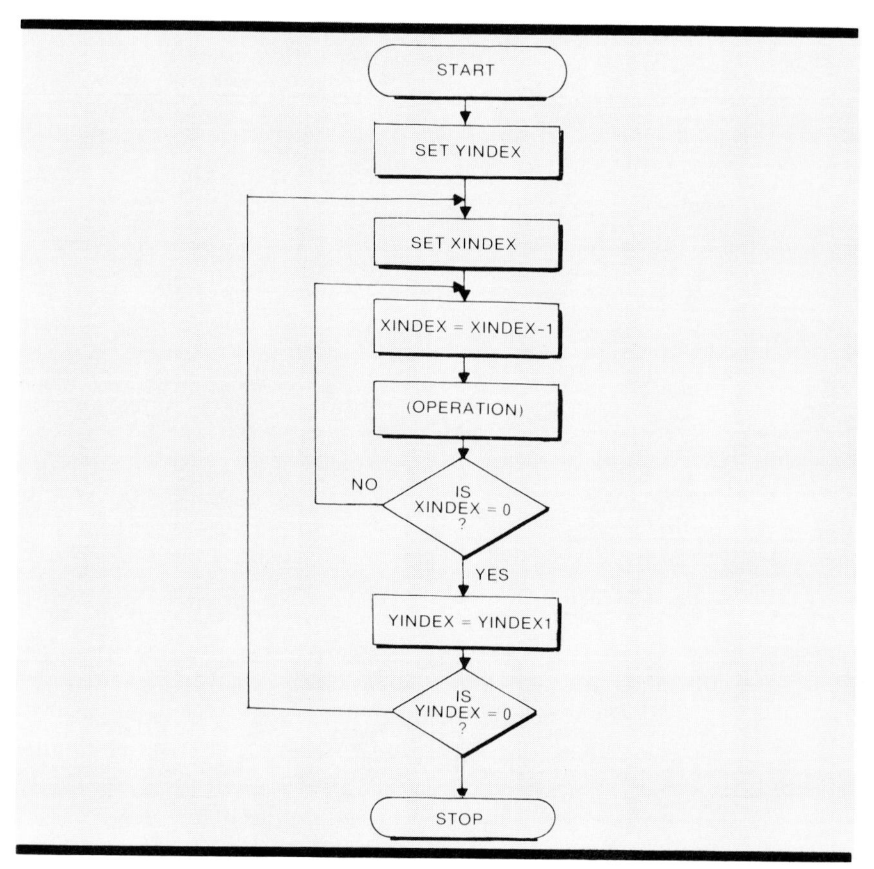

Fig. 8-2. Loop Index Program

8-6. A Peripheral Control Program

The last program we will discuss in this unit is a segment of a control program driving a printer from a microprocessor. This segment determines whether a character input from the keyboard is to be printed as a character or is actually an "escape code" character which is to be used for some printer-control purpose, such as a formfeed or change in line spacing. If it is an escape character, it is sent to the printer but not to the CRT monitor, where it may have some conflicting meaning, e.g., it could blank out the CRT. Instead, the CRT monitor is sent a dollar sign "$" to display the fact that a control character was sent out.

Example 8-6
(See Table 8-2)

```
0050TOUT       .DE $8B86        ;OUTPUT TO CRT
•
•
•
0230PRINT      CMP #$1B         ;COMPARE WITH ESCAPE
0240           BNE NOESC        ;BRANCH IF NOT ESCAPE
0250           JSR WAIT         ;GO TO WAIT ROUTINE
0260           LDA #'$          ;PUT "$" IN ACCUM
0270           JSR TOUT         ;OUTPUT $ TO CRT
0280           RTS              ;RETURN
•
•
•
0620WAIT       PHA              ;SAVE ACCUMULATOR
0630WAIT1      LDA $A801        ;GET CONTROL BYTE
0640           ROR A            ;MOVE BIT 1 TO CARRY
0650           BCS WAIT1        ;WAIT IF BIT NOT SET
0660           PLA              ;POP ACCUM FROM STACK
0670           RTS              ;RETURN FROM SUBROUTINE
```

TABLE 8-2. Printer Control Program Segment

At line 50, TOUT is the label for a monitor subroutine that outputs the accumulator contents to the CRT display. The address $8B86 within the monitor is designated as an "external" address to the assembler by the pseudo op .DE.

When the PRINT subroutine is called, the character in the accumulator is compared with the value of the escape code $1B. The CMP (Compare Memory and accumulator) instruction is similar to a subtraction operation, but does not actually change the value in the accumulator. However, it does affect the zero flag in the same way — Z is set if the two are equal and the result of the subtraction is therefore zero. (This might be considered a "nondestructive test" of equality since it does not destroy the value in the accumulator as a subtraction would.) If the two are not equal, Z=0 and BNE causes a branch to the location labeled NOESC, where the byte is tested for a number of other alternatives not considered in this segment.

If the byte is an escape code ($1B), the program jumps to the subroutine WAIT at line 620. The purpose of this subroutine is to wait until the printer is ready to accept the new input (it may be working on another character). We can determine if it is ready by looking at the bit 0 of location $A801, which is an address in a peripheral control chip wired directly to the printer. If this bit is not set, the printer is ready. (This is called a *handshake*.) Therefore, the subroutine pushes onto the stack the current value of the accumulator and gets the control byte at $A801. The ROR instruction (Rotate Right) brings the LSB into the Carry. If LSB was set, then Carry is now also set, and the instruction BCS causes a branch back to WAIT1. This loop repeats itself until the bit is not set. At this time, it pops the byte for the printer back on the accumulator and sends it to $A800, which is the printer wired interface.

The RTS returns control from the subroutine to line 260, which loads a dollar sign (using the RAE convention '$ to represent the ASCII hex value of "$") on to the accumulator. This is sent to TOUT where it is caused to appear on the monitor.

This program segment demonstrates the use of branches, subroutines, and loops in a manner typical of most assembly programs. It is worthwhile going back over this explanation and the table several times until you are sure you understand it. You will then be in a position to read other assembly language programs and find out what they are trying to do.

Exercises

8-1. *A line of code is written as follows:*

100START LDA TABLE+1 ;SINE TABLE

After assembly it appears as:

0500– AD 01 50 100START LDA TABLE+1 ;SINE TABLE

a) What is the line number?
b) What is the label?
c) At what addresses are the instructions and operand?
d) What is the address of TABLE+1?
e) What does the assembler do with the words "SINE TABLE"?

8-2. a) Write the RAE assembly-language source code for a program to move data from memory location $2010 to location $1000. Use 6502 instructions in Appendix A. (Omit pseudo ops.)
b) Manually assemble your program into object code beginning at $0500 and write the hex dump, including checksums.

8-3. In Example 8-2 change the initial contents of $500 to FF. Write the object code.
a) What appears in location $600? Is it correct?
b) How can you detect an error? (Hint: test a flag.)

8-4. In Example 8-2 change the location of the addend data from $10 to $400.
a) Write source and object code.
b) What is the difference in execution time (clock cycles)?

8-5. a) Write the object code for Example 8-3.
b) How many additional instructions would be required to add two 24-bit numbers? What would they be?
c) Would it make any difference in the result if the CLC was the first instead of the second instruction? If it were after the ADC?

8-6. What is the greatest delay possible with the scheme of Example 8-4?

8-7. What modification would make Example 8-4 into a subroutine so that it could be used in other programs?

Unit 9: BASIC High-Level Programming

Unit 9

BASIC High-Level Programming

In the previous unit you learned the fundamentals of assembly and machine language programming. Each type of processor has an assembly language as unique as its instruction set, but you have seen enough of one language to capture the flavor of the others. Most people find it easier to learn and use high-level languages because they are oriented toward the problems to be solved with the computer and are designed to interface with the human as well as the machine. This is especially true of application engineers; people like yourself, probably, who are concerned primarily with using the MPU as a control tool, not with the mass production of a computer game or household appliance. This unit will introduce you to high-level languages to demonstrate their features and advantages, and to motivate you to learn more so that you can apply them to your problems.

Learning Objectives—When you have completed this unit you should:

A. **Recognize the structure and style of programs designed in the BASIC language.**

B. **Understand some of the instructions and programming constructs in this language and how they are used.**

C. **Be able to compare the usefulness of this language with assembly.**

9-1. BASIC as a Control Language

FORTRAN is probably the oldest and most widely used programming language for engineering problems, since it is taught almost universally in engineering schools. But it is poorly suited to microprocessors, because of its large compiler and lack of more modern structural programming concepts. BASIC is only slightly newer and no better structured, but it is a very popular microcomputer language since it is comparatively easy to learn and uses a relatively small interpreter.

Aside from its popularity for personal computers, enhanced forms of BASIC are making inroads into the control field. Some versions that are commercially available are XY BASIC (Mark Williams Co., Chicago), Power BASIC (Texas Instruments, Inc.), and Process BASIC (PBASIC) FROM Mostek Corp. (Ref. 1). PBASIC, for example, is capable of performing PID calculations, formating data reports, and manipulating bits for event control functions, working with a real-time clock. It is available as an interpreter on a ROM chip and in conjunction with a hardware mathematics board can operate at about the same speed as standard BASIC interpreters, although only half as fast as a BASIC compiler such as Microsoft's.

The European Computer Manufacturing Association and the American National Standards Institute (ANSI) are preparing an Industrial Real-Time BASIC (IRTB) standard (Ref. 2) that has specific features for the concurrent real-time activities encountered in distributed control systems, such as plant interface input/output, messages, and shared database management. IRTB, when available, should do much to popularize the use of BASIC as an alternative control language.

9-2. Search Program Example Algorithm

(This example and the remainder of this unit are extracted from the author's book *Microprocessor Systems, Interfacing and Applications* (Ref. 3). For a fuller explanation of the concepts discussed, the reader is referred to this source, and to the many textbooks written about the BASIC language.)

The program examined is one often used as a "benchmark" test to compare microprocessor performance, called "table search" or "table lookup" (Ref. 3). Let us assume we have a table of N numbers placed in a block of sequential memory locations, but the data in this table is not in any particular order. We will want to search this table to find the presence, and location if it is present, of a specific number, X, called the key, which we furnish as an input to the program. If X is present, the program will output its location (its order on the list), denoted j, while if it is not present, the program will output "0" after completing the search.

In the "top-down" tradition, this program can be represented by an overview in the forms of a flowchart, Fig. 9-1. (Recall Sec. 7-1 and Fig. 7-2 for the meaning of block symbol shapes.) The flowchart portrays graphically what we have just expressed in prose: we should input the number X to be searched for, compare it with all the table entries, output its location if KEY is found, etc. But the flowchart does not say how we should search the table, although there may be many ways of doing this. For example, we can look at each location in sequence, or we can search them in some pattern, such as even first and then odd. Each method results in a different algorithm (see Example 7-1) and presumably they will differ in important respects, such as speed of execution, memory space required, and so on.

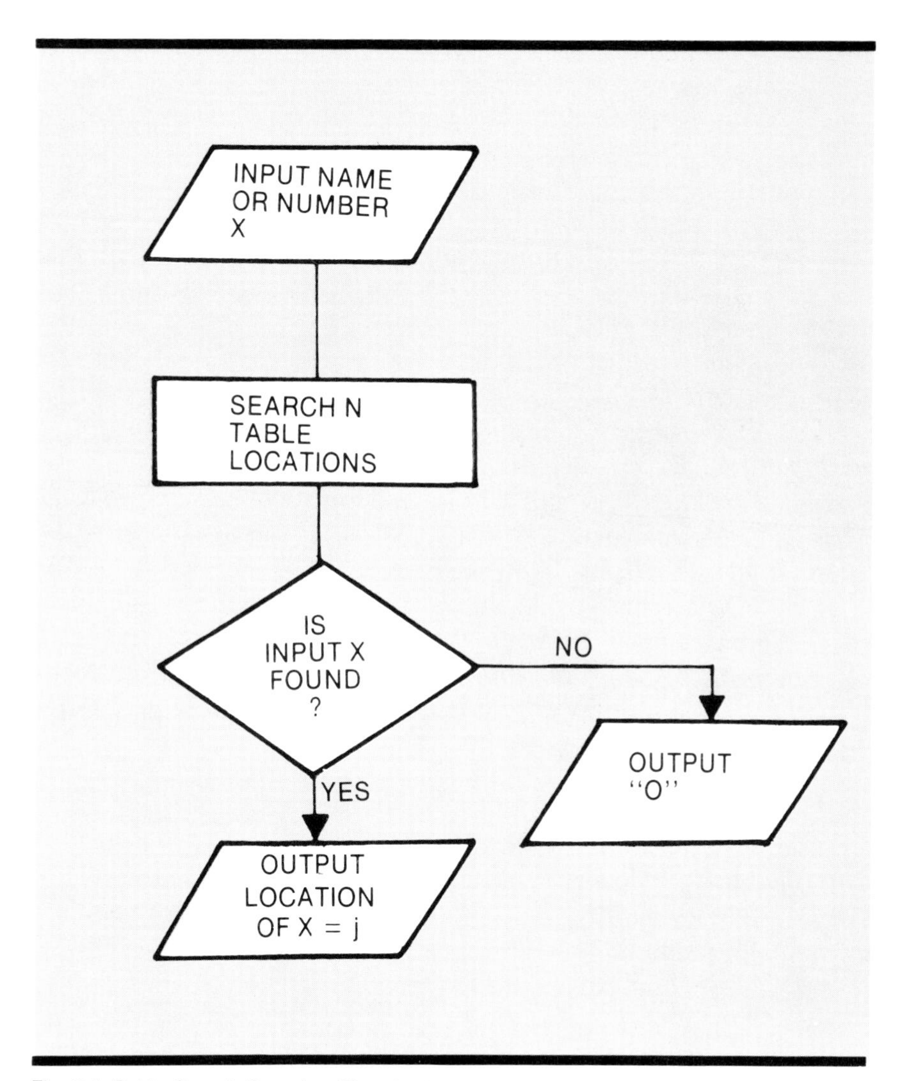

Fig. 9-1. Table Search Overview Flowchart.

Note also that this task need not be trivial, since the numbers may represent some significant items, such as stock in an inventory or people on a payroll list, or a mathematical table. The ability to detect location or absence of any key item is the nucleus of an information retrieval system.

Donald E. Knuth, a professor of computer science at Stanford University, has discussed at length various algorithms for searching lists in a number of books and articles, of which Ref. 4 (in the *Scientific American* magazine) is highly recommended. For example, in a list in which table entries are previously sorted by size, a method called *binary search* is very efficient, since the key need only be compared with the boundaries of successive halvings of the list. In a list of 10,000 items, the average requirement is 5000 comparisons to find a key item if a "brute force" search is employed, that is, the examination of each item in sequence until the key is found. Only 14 comparisons are needed for a binary search.

However, for this example we will use Knuth's "brute force" algorithm since it is simpler to explain and doesn't need the sorting step.

Example 9-1
Table Search Algorithm (Ref. 4)

1. (Initialize.) Set j equal to N (Remarks: N is the number of words in the sequential table. We assume that it is also the address of the last word in the table. The algorithm will search each address from N to 0 and output 0 unless it finds the input word X first, in which case it will output the address j where X was found. Note: In Fig. 9-2 the backward arrow means "set equal to." This symbol avoids confusion with the equals sign "=" which tests for equality.)

2. Unsuccessful? If j=0, output j and terminate the algorithm. (Otherwise continue to the next step.)

3. Successful? If X = KEY(j), output j and terminate the algorithm. (Otherwise continue to the next step.)

4. Repeat. Set j equal to j—1 (that is, decrease the current value of j by 1) and go back to Step 2.

This algorithm is flowcharted in Fig. 9-2. The logic is laid out very clearly in this graph. It can be seen by following through the sequence indicated by arrows that all requirements of the algorithm are met. The algorithm steps down each address searching for X in KEY(j) and when it is found, outputs j and stops. Whenever the end of the table is found without finding X, the value j = 0 is output.

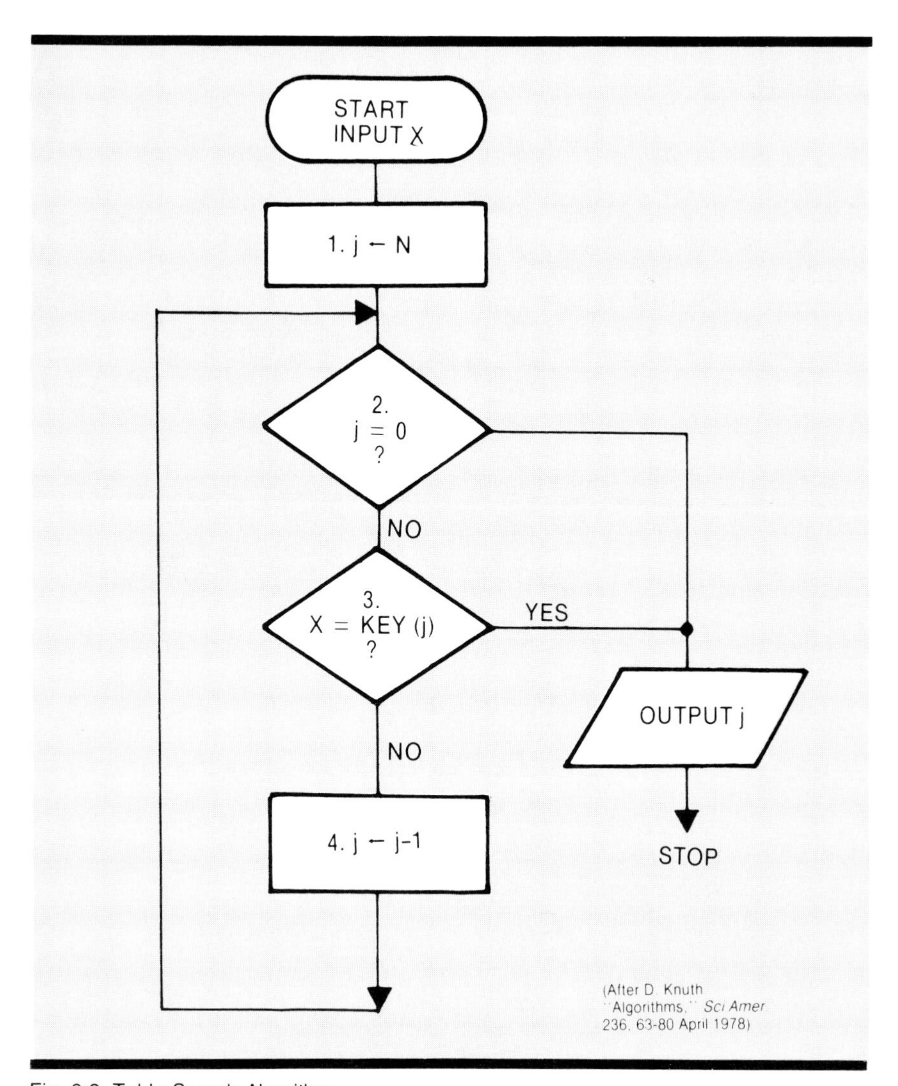

Fig. 9-2. Table Search Algorithm

9-3. BASIC Search Program

BASIC is well matched to human thinking and can be characterized as having an almost English-like structure that is nearly immediately understood by anyone familiar with that language and algebra.

Perhaps the best way to demonstrate this is to go through an actual program and point out the features as they occur. Consequently, we will continue with the search algorithm just developed (Fig. 9-2). In BASIC, the flowchart can be coded almost immediately.

To be concrete, assume we have 10 numbers on the list, which will be the ASCII codes for the first initials of 10 names. The list will look like Table 9-1.

Address	Name	ASCII Code
1	JONES	74
2	SMITH	83
3	BROWN	66
4	CLARK	67
5	WILLIAMS	87
6	ABLES	65
7	THOMAS	84
8	PERKINS	80
9	LALLY	70

TABLE 9-1. Data to Be Searched

Using these data, the algorithm of Fig. 9-2 can be written as a BASIC program, Table 9-2.
Note that each line of the program is give a line number, much as in assembly programming.

```
1 REM THIS PROGRAM SEARCHES A LIST AND PRINTS THE ADDRESS
10 DATA 74,83,66,67,87,65,84,80,76,70
20 LET N = 10
30 LET X = 65
40 LET J = N
50 IF J = 0 GO TO 100
60 READ K
70 IF K = X GO TO 100
80 LET J = J—1
90 GO TO 50
100 PRINT J
110 END
```

TABLE 9-2. Basic List Search Program

Already this program should make a certain amount of sense. After a REM (remark) line explaining what the program is to do in plain language (ignored by the BASIC interpreter) we start at line 10 with a list, which is obviously the name codes we are about to search through. Appropriately, this line is called DATA. On line 20, we set forth the value of N, the number of items in the list. X, on line 30, is the code item for which we are searching, the ASCII decimal code for the letter A.

On line 40 we begin the algorithm of Fig. 9-2 by setting J equal to N, or 10 in the example. If, in line 50, J = 0, we "GO TO" (jump) to line 100. Looking ahead, we see that line 100 commands us to PRINT J. This is in accordance with Step 2 of the algorithm. If J does not equal 0, we read the next line (60) which commands the computer to READ K, where K stands for KEY(j) of the algorithm, Step 3. In BASIC, the command READ means to read the DATA list items in sequence, one for each READ. Thus, the machine will read the number first on the DATA list, or 74. (The next time it encounters READ it will choose 83, having "used up" 74.)

The value read from the DATA list is assigned to the variable K. Line 70 states that we should GO TO 100 if K = X. Since K now equals 74 and X was originally given the value 65, the statement K = X is not true. The rule for IF statements is that they are ignored if not true. Hence, we do *not* jump to 100 but instead read the next line. Line 80 is the same as Step 4 of the algorithm; we assign to J a value that is one less than its present value. So now J = (10-1) or J = 9.

The astute reader will note that we have used the equal sign "=" in two different ways. In lines 80 and 20 through 40 the "=" following LET means "assign the value of the *expression* on the right to the *variable* on the left. In this case the arithmetic expression (J-1), where J = 10, is evaluated, resulting in a value of 9 which is then assigned to the variable J. In lines 20 through 40, the "expression" is merely a constant, 10 or 65. (Note that in many versions of BASIC the LET is optional, and line 80 could have been written, for example,

J = J − 1

LET is then often reserved as a means of "flagging" new variables entering the program.

The second meaning of the "=" sign follows the IF statement. This is used more like the conventional "=" of algebra. In the IF statement we are interested in the *relation* between the two expressions following IF, whether it is true or false. In the statement "K = X" K and X can be any expressions, and the relational operator is "=." If K does indeed equal X, then the Boolean value of the statement "K = X" is TRUE. Otherwise it is FALSE. In the IF—THEN statement, or the equivalent IF—GOTO of lines 50 and 70, the statement following THEN (or the GOTO) is obeyed only if the Boolean value following IF is TRUE. Otherwise, it is ignored and the next line read.

After decrementing J in line 80 we arrive at the next line which is an *unconditional branch*, GO TO (line) 50. This follows the algorithm which states that the decrement of address J is followed by retesting to see if we have yet reached the end of the list (J=0) without a "hit." The program, following the algorithm, tells us to print "0" in that event.

In the example we continue to shuttle between lines 50 to 90 and back, reducing the address J by one each time and testing to see if K becomes 65. The "hit" is reached when the value of J reaches 5. Then the relational statement of line 70 becomes TRUE, we GO TO 100, and print the current value of J. Thus, the result of the entire program is merely the printout.

5

meaning that the key word was found in location 5.

9-4. Additional Rules for BASIC

The above explanation treated BASIC in a somewhat wordy and intuitive manner; actually the rules of *syntax* of this, or any programming language, must be stated precisely and compactly. Unfortunately the syntax definitions are more difficult to learn than the language itself unless you are trained in logic and symbolic algebra. We will not go into this strict representation here, but will restate the BASIC rules applying to this program with more rigor.

LINE NUMBERS: Every line of BASIC must be preceded by a line number followed by a space. Only integers are permitted as line numbers. Zero is not a valid line number but leading zeros (such as 007) usually are permitted. There is some upper limit on line numbers, depending on the BASIC dialect; for example, 1 to 9999.

Line numbers need not be consecutive but statements must be uniquely numbered in ascending order. However, lines may be entered in any order and the interpreter will automatically order the file in ascending sequence. It is good programming practice to number lines in multiples of 10, e.g., 100...110...120...; then if one must insert a new line between two existing lines there is plenty of space. If a line number is duplicated, BASIC will delete the earlier line and substitute the new one; this feature may be used for editing and correction. If only the line number, followed by carriage return (CR), is entered, that line is deleted.

Line numbers (ln henceforth) are also used as *statement markers*. In the example program, GO TO 100 uses the ln 100 as a guidepost to transfer control to the statement line 100.

SPACES: Following ln there must be (in most BASICs) one blank character (space). The remainder of the statement line may or may not contain spaces. The spaces are ignored by the interpreter, but may be inserted to make the program easier to read.

REMARKS: Any statement on a line beginning with REM is ignored by the interpreter and may be used by the programmer to explain the purpose or method of the program. These remarks serve to remind the programmer or some other person who may want to make some revision in the future what was meant by the original coding. Some BASIC variations use the single quote (') as an abbreviation for REM, or to add a remark to a line that contains an active statement, for example:

```
100 END 'THIS IS THE END OF THE PROGRAM'
```

DATA: The DATA statement has already been discussed. DATA lines may be put anywhere in a program and form a *pool* of data which can be READ by the program. Some BASICs allow *strings*, groups of ASCII characters, to be put into a DATA pool. The word RESTORE allows the DATA pool to be reused, starting at the first DATA item.

CONSTANTS: A constant is a signed number. Small BASIC compilers may be limited to integers. If there is no sign, the constant is assumed to be positive. A floating point number, if allowed, is distinguished by a decimal point. Some interpreters allow exponential numbers (to the base 10) to be entered, using E as the "exponent of 10" symbol.

5.65E6 means 5650000
0.3E−2 means .003

Constants keep their assigned values throughout a program.

VARIABLES: Variables are those quantities that may change in value during the program, such as J and K of our previous example. BASIC allows any letter to be used as a variable name, or a letter followed by a single digit, 0 to 9 (or another letter). Thus, legal variable names are A, X, N, or A1, A2, X9, but C3P0 would be rejected (or interpreted as C3), while 3J is illegal since it starts with a number, not a letter. A variable may take on any value that is permitted for a constant. A variable does not have to change value during a program, but is a convenient way to label inputs, and permits them to be changed if the program data should be changed. For example, in our search program, N was not changed from its initial value of 10, but if the number of data items were to be altered, N can be given a different value.

EXPRESSIONS: Variables and constants can be combined with arithmetic symbols to form algebra-like formulas called BASIC *expressions*. J – 1 in line 80 is an example. Another is:

    3*A – 1.6

where * is the multiplication operator and +, – have their usual arithmetic meanings. Parentheses may be used, as in:

    2*(3+X)

12 in which case the expression within the parentheses is evaluated first, and the remaining operations conducted in order of the priority shown below.

"up arrow" or **	exponential
*	multiplication or
/	division
+	addition or
–	subtraction

The rules of forming expressions are:

1. Operators may not be adjacent; X*+2 is illegal but X*(+2) is alright since the parenthesis separates the operators.
2. Variables and constants may not be adjacent. 2X is illegal but 2*X is alright.
3. Parentheses must enclose a legal expression. They may be nested as in A*(B+C*(D–F)).

Note that for every opening (left) parenthesis there must be a closing (right) one. It is good idea to count them in a complex expression to be sure that they are all paired.

Following precise rules when evaluating expression avoids some of the ambiguity found in ordinary arthmetic. For example, what is the value of

30/3*2 ?

It may be either 20 or 5 depending on which operator is applied first, the division (/) or multiplication (*). In BASIC the rules are:

1. Expressions in parentheses are evaluated first, starting with the innermost pair.
2. Expressions are evaluated in order of the priority of the operator, as noted in the table presented above.
3. When priorities are equal, evaluate from left to right.

9-5. Calculation and Program Control

The LET statement is the simplest arithmetic assignment statement. It is of the form:

```
   ln LET v = e
or ln v = e
```

where v is any variable name and e is any arithmetic expression (including a constant). The equals sign "=" means "be assigned the value of" or "be replaced by."

Note that BASIC can be used as a kind of calculator by using the statement:

```
PRINT e
```

where e may be any kind of complex expression. A line number is not used in this mode; the expression will be evaluated and printed out but will not be saved or stored for any other use. The simplest *control* statement is GO TO or GOTO, an unconditional branch to the indicated line number:

ln GOTO ln'

GOTO controls the program by altering the normal sequence of execution in order of line number, by forcing a branch to ln'. GOTO can be used to program a *loop*, a repeating series of commands, as was done in our sample program. The loop in infinite, e.g., will go on forever, unless an escape statement is included to get out of the loop, such as lines 50 and 70 in the example.

The IF THEN (IF GOTO) statements, such as lines 50 and 70, rely on something changing within the program so as to make them become TRUE or FALSE. The "something" is the relational operator following IF, of the form:

e(1) op e(2)

where e(1) and e(2) are any expressions and the operator "op" is one of the following list:

Operator	Meaning
=	equal to
< = or = <	less than or equal
<	less than
>	greater than
> = or = >	greater than or equal
< > or > <	not equal to

Thus, 2*A> = 8 is TRUE if the variable A takes on the value of 4 or more during the program, otherwise it is false.

The statements:

ln IF e(1) op e(2) THEN ln'

ln IF e(1) op e(2) GOTO ln'

have the identical meaning, a *conditional transfer of control* to line number ln' if the relation e(1) op e(2) is TRUE. Another form of IF statement found in some BASIC interpreters is:

ln IF e(1) op e(2) LET e(3) = e(4) or

ln IF e(1) op e(2) THEN e(3) = e(4)

This substitutes an arithmetic expression for the control statement.

Example 9-2

```
100 IF A > B LET X=15
```

assigns the value 15 to the variable X if and only if the variable A is greater than the variable B.

9-6. Input/Output

One of the nice features of BASIC for the tyro programmer is that you do not have to worry about the format of printing your output, unless you want to do so. In FORTRAN, and even more so in COBOL, a large fraction of the programmer's effort is taken up by specifying the exact output format to follow. BASIC will automatically give a simple but usable output on the bare statement:

```
ln PRINT a(1), a(2), ...a(n)
```

where the a's are any items to be printed. If they are separated by commas, a nominal zone, usually 13 spaces, is reserved for each item and they are printed right-justified within that zone.

If semicolons are used in lieu of commas the items printed will be more densely packed with only enough space to separate them. The a's may include variables, in which case the current value is printed, or constants, or even expressions. The current value of variables or expressions is printed out as in the "calculator" mode.

We also can print out what is called an alphanumeric *literal* or a string of literals. A literal string is any character or character sequence (including spaces) that is enclosed in quotation marks, for example:

```
10 PRINT
20 PRINT"THIS PROGRAM SEARCHES LISTS"
```

The first statement (10) produces a blank line while the second statement prints the characters within the quotes, followed by C/R.

Input has been considered in our sample program only through the method of constants in a DATA list. This is not the best way to introduce data because the program line must be changed each time the list is altered. Data also can be entered from a terminal keyboard using the statement INPUT.

ln INPUT v(1), v(2), ... v(n)

where the v's are variables.

The program will halt on encountering INPUT and print out

?

which should be followed by a constant value entered from the keyboard by the operator.

Example 9-3

```
10 INPUT X
20 PRINT X + 2

? 10 (operator enters underlined number)
12 (computer printout)
```

More than one constant can be entered after the question mark; these are separated by commas and terminated by a C/R. Values are assigned to each of the variables on the list in sequence. Extra constants are ignored, and if not enough are supplied, another ? will be printed

Some BASICs also allow a quoted literal to be printed with INPUT, such as:

100 INPUT "ENTER VALUE FOR Y"; Y

9-7. FOR-TO-NEXT Loops

These constructs permit looping or repetition in a more sophisticated way than GOTO or IF THEN statements. The FOR-TO-NEXT statement has three parts, a counter variable, an end test, and an increment or step. The general statement is 12

ln FOR v = e(1) TO e(2) STEP e(3)

where v is the counter variable, e(1) is an expression which when evaluated becomes the initial value of v, e(2) is the final value of v, and e(3) is the increment added to v at each repetition. If STEP is omitted, e(3) is assumed to be 1. A block of statements which are to be repeated follow the FOR line. After the end of the block is the NEXT statement:

```
ln NEXT v
```

where v is the same variable chosen as a counter.

Example 9-4

```
10 FOR X = 1 TO 9 STEP 2
20 Y = 2*X
30 PRINT Y;
40 NEXT X
```

will result in the printout

```
2 6 10 14 18
```

The block of statement 20 through 30 will be executed first for X=1, then X=1 + 2 or 3, X= 5, etc., adding the STEP value of 2 to X cumulatively until the value 9 is reached. Since another increment would exceed the upper limit of 9, the statement terminates and continues to the next line in the program, if any.

Example 9-5

If STEP 2 is omitted in line 10 above, the increment is assumed to be 1, and the printout is:

```
2 4 6 8 ... 18
```

The value of STEP can be negative (counting backwards) or even a floating-point value such as .5 in some BASICs.

9-8. Other Variations of BASIC

Most of the BASIC features mentioned here are supported by the ANSI standard. Some smaller interpreters do not support all these features, especially those of less than 4K in size. Some of these are confined to integers between 0 and 255. Other features, supported by the larger interpreters and compilers include the following:

ATN(n): returns arctangent (radian value)
SIN(n), COS(n), TAN(n): trig functions
LOG(n) and CLOG(n): natural and common logs
SQR(n): square
DIM(n,m): a method of reserving space to store element of arrays or matrices (2 dimensional arrays)
ELSE: alternate branch in an IF statement
ON GOTO: a three-way branch statement
FN: allows user-defined function to be named
GO SUB: branch to a subroutine. RETURN goes back to main program

Any BASIC interpreter or compiler will be furnished with documentation that is the final authority for that version of the language. A good reference to the many variants is Ref. 5 by David Lien.

Exercises

9-1. *Evaluate the following BASIC expressions:*

```
(4*3)*2/( (5+1)*2 )
```

9-2. *a) In the BASIC program of Table 9-2 what changes would be made to what line(s) if the name to be searched for was BROWN. Hint: recall that the ASCII code for the initial letter was used.*
b) What would be the result of omitting line 50?
i) if x=65 ?
ii) if x= 77 ?

9-3. *Using the concepts of Table 9-2, write a BASIC program to print squares of numbers from 1 to 9 inclusive. Hint: PRINT N*N will print the square of N.*

9-4. *Given the program is BASIC:*

```
10 X=2
12 IF X=3 GOTO 16
14 PRINT"TRUE"
16 PRINT"FALSE"
```

What will the program do?

9-5. *Evaluate the BASIC following expressions:*

```
(3*A−1.6) for A=2
2*A**2+3*A+4 for A=2
3*(4−16/2)−3*(2**2−1)
2*(3+4*(5−2) )
```

9-6. *Using FOR NEXT STEP, write a program to print the cubes of even numbers from 2 through 10.*

References

1. Chaky, M. et al, "Process Control BASIC Simplifies Programming," *Instruments and Control Systems*, January 1982, pp. 51-54.
2. "Industrial Real-Time BASIC (EWICS TC2 81/8), Draft Standard European Workshop on Industrial Control Systems, c/o A. Lewis, AERE Harwell, Oxon, England.
3. Bibbero, R.J. and Stern, D.M., *Microprocessor Systems, Interfacing and Applications*, New York: John Wiley & Sons, (1982), pp. 86-88.
4. Knuth, D.E., "Algorithms," *Scientific American*, April 1977, pp. 63-80.
5. Lien, D.L., *The BASIC Handbook*, San Diego: Compusoft (1978).

Unit 10: Programming in FORTH

Unit 10

Programming in FORTH

FORTH is not an easy language to learn or understand but has become important as a microprocessor control-system language because of its speed and capabilities. It is our intention in this unit to expose you to FORTH concepts and to a program similar to the BASIC program examined in Unit 9. This will give you a background to help decide whether to use FORTH for your control problems.

Learning Objectives—When you have completed this unit you should:

A. Recognize the features of the FORTH language, its concepts of programming, and the method of compiling a dictionary of FORTH words.

B. Understand the workings of a program in FORTH.

C. Know how to write elementary FORTH program segments and dictionary definitions.

10-1. Fundamental Features of FORTH

In Sec. 7-2 we described FORTH briefly as a high-level language using a *threaded incremental compiler. Threaded* refers to the fact that a dictionary of FORTH words is defined by using previously defined FORTH words until they lead (thread) back to a small core of basic operations couched in machine language. The *incremental* part means that each definition is compiled as it is defined, building up the dictionary for a specific job. The process of FORTH programming is the act of defining specialized words to perform your program. Because each program need use only its specific definitions (and their previously defined root words) the vocabulary for any particular program can be kept small and specialized. Since the words are compiled, they run faster than they would under an interpreter language like BASIC. FORTH runs about ten times as fast as BASIC; in fact, it is close to machine language in speed. The small vocabulary means that little memory space is needed to run a particular program. The combination of speed and economical memory results in a language that is very good for control problems. In fact, FORTH was developed to solve one of the most difficult control problems on earth, the stabilization of the multiple-mirror telescopes at Kitt's Peak Observatory in Arizona, a job that involves coordinating mechanically separate parts of the telescope within a fraction of a light wavelength and at speeds faster than the fluctuation of the atmosphere's turbulence.

In spite of FORTH's machine language-like performance, it is capable of being used in a very sophisticated way and uses advanced programming constructs and structures, actually more advanced than most versions of BASIC. These constructs include the IF-ELSE-THEN and BEGIN-WHILE as well as the DO-LOOP which is similar to the BASIC DO-NEXT.

These are the virtues of FORTH. The problem in using it is that it requires a different way of thinking. In fact, FORTH turns your 8-bit microcomputer into a 16-bit stack machine. A stack machine does its operations in post-fix or Reverse Polish Notation (RPN), as explained in Sec. 6-9. Consequently, to read and write FORTH programs we must learn to think and calculate in RPN style, rather than the normal algebraic operations we are accustomed to using. We also must think about "the stack" because all operations consist of manipulating numbers that are on top of the stack, so that it is necessary when programming to always be aware of what is on the stack, what operations put numbers on it, and what removes them. (Actually, there are two stacks, but we will consider only one here since the other is used only by the machine and by advanced programmers.)

10-2. Plan of This Unit

It is well beyond the goals of this unit to teach FORTH programming. This would require at *least* an entire book, of which several are recommended (Refs. 1, 2, 3). Rather, we will transmit the flavor of the language by examining one program. This program performs the same function as the BASIC code studied in the last unit, that of searching a list for a key number. Consequently, you will have an opportunity to compare two languages performing the same task.

Before going on to explain the list search program, we will briefly review post-fix operations as they apply to FORTH and study the method of dictionary definition.

10-3. Stack Operations

Fundamentally, FORTH operates on 16-bit integers, not real numbers with decimal points. This is one of its weak points. Integer numbers are placed on the stack merely by inputting them, separated by spaces, from the terminal or other input device. The last number entered is on top of the stack. They can be removed from the stack, top first, and displayed on the screen by using the FORTH word "period" (.). (FORTH words may look like punctuation marks or whatever. Whenever confusion might exist between a FORTH word and some other meaning, parentheses ( ) will enclose the word.) Computer responses to FORTH word executions are underlined.

There is only one real punctuation rule in FORTH: all FORTH words must be separated by at least one space.

Example 10-1

Type in:

10 11 32000 5 (ret.)

There are four numbers on the stack, the top is 5 and the bottom is 10.

Entering: . . . . (ret.)
will product: 5 32000 11 10

(the top of the stack comes off first).

Various FORTH words manipulate the items on the stack.

The word SWAP will reverse the two top stack items:

5 10 11 13 SWAP . . . (ret.)

will print out

11 13 10 (note we left 5 on top of the stack).

DUP duplicates the top of the stack and DUP? does so only if the top is non-zero.

Entering 5 10 11 13 DUP DUP

results in the stack containing:

BOT> 5 10 11 13 13 13 <TOP

(The stack display above can be obtained by the FORTH word (S.) which is a "nondestructive stack print." Unlike (.) it does not remove anything from the stack.)

Likewise, DROP throws away the top of the stack, OVER copies the second item from the top and puts the copy on top, ROT rotates the third entry to the top of the stack, so that

10 11 13 5 ROT S. (ret.) results in:
BOT> 10 13 5 11 <TOP

These and other stack manipulation words permit any item to be moved within the stack.

10-4. FORTH Arithmetic

The basic arithmetic operations hold in FORTH except that the arithmetic is integer, that is, ordinary division will show no remainder. (However, there is a FORTH operation MOD, pronounced modulo, that will preserve the remainder on division of n1/n2 on top of the stack, with the sign of n1.) The operation + − * and / perform respectively:

+ remove top two from stack, add, replace sum on top

− subtract top from top−1, replace difference on top

* multiply top by top−1, replace with product on top

/ divide top into top−1, discard remainder, replace quotient on top.

Example 10-2

5 10 + . (ret.) 15
6 17 − . (ret.) −11
6 8 * . (ret.) 48
45 13 / . (ret.) 3

FORTH also will handle double precision (32-bit) numbers and has a special set of words for these operations.

10-5. Memory Operations, Constants, and Variables

An integer constant can be named by the FORTH word CONSTANT. Henceforth, calling the name will place the value of the constant on the stack. This is good programming practice since the name may be more meaningful than the number.

Example 10-3

31416 CONSTANT PI (ret.)

creates a dictionary item named PI with the value 31416. To retrieve the number from its assigned memory location it can be placed on the stack by calling its name.

PI . (ret.) 31416

The value of the constant cannot be easily changed once set.

If a value is to be varied during a program, the proper construct is VARIABLE. A variable is a name for a memory location, as is CONSTANT, but the value stored in VARIABLE can be easily changed. In some versions of FORTH (2) a variable is declared with an initial value, but in standard FORTH the variable is declared first and then given a value (3). For example, declaring

VARIABLE RADIUS

creates an address with the name RADIUS. Executing RADIUS will not retrieve the value but will put the *address* on the stack. To give RADIUS a 16-bit integer value we use the FORTH word (!), pronounced "store," as in the following code:

5 RADIUS !

In order to retrieve the value we just stored in the variable, we must first put the address on top of the stack by executing the word RADIUS and then fetch the contents of that address by using the word (@).

RADIUS @. (ret.) 5

Example 10-4

(Note: The word (*/), called "times divide," operating on the stack

BOT> ... n1 n2 n3 <TOP

will multiply n1 by n2 using double precision arithmetic and then divide by n3, leaving only the integer value of the qoutient. This is a more precise operation than separate * / .)

RADIUS @ DUP * 10000 */ . (ret.) <u>78</u>

will multiply n1 by n2 using double precision arithmetic and then divide by n3, leaving only the integer value of the quotient. This is a more precise operation than separate * / .)

RADIUS @ DUP * 10000 */ . (ret.) <u>78</u>

10-6. Character Strings and Formatting

Messages in the form of strings of ASCII characters can be output from the terminal by the word (.“), called "print message." This command prints on the screen whatever follows between the double quote and the next double quote (”) encountered. Thus,

.“ THIS IS A MESSAGE ”

will print out

THIS IS A MESSAGE

on the monitor screen.

To start a new line, use the word CR (for carriage return). SPACE or SPACES will insert the number of spaces specified by the top of the stack.

Example 10-5

CR.“ COL 1 ” 5 SPACES .“ COL 2 ” (ret.)

will produce the heading

COL 1   COL 2

on the screen.

10-7. FORTH Conditional Structures

FORTH contains program-control constructs like the IF-THEN conditional and the FOR-TO-NEXT loop of BASIC. The simplest conditional is IF-THEN. The construction is:

... flag IF .. (execute if flag is TRUE) .. THEN ...

If the top of the stack when IF is encountered is a TRUE flag all words between IF and THEN are executed. If the flag is FALSE the program skips directly to the words following THEN. A 1 or any 16-bit number other than 0 on top of the stack is considered a TRUE flag; 0 is FALSE. The flags are usually left on top of the stack as the result of a previous *comparison operator*. The comparison operators for signed numbers are < = > 0< 0= 0> and NOT. These have the following meanings:

Stack	Operator	Flag
n1 n2	>	TRUE if n1 less than n2
n1 n2	=	TRUE if n1 equals n2
n1 n2	>	TRUE if n1 greater than n2
n	0<	TRUE if n less than 0
n	0=	TRUE if n equals 0
n	0>	TRUE is n greater than 0
n	NOT	TRUE if n = 0 (reverses previous value of flag)

Table 10-1. Comparison Operators

Example 10-6

5 10 < . (ret.)1

(n1, n2 removed from stack, replaced with TRUE flag = 1)

9 9 = . (ret.) 1
9 5 = . (ret.) 0

(the carriage return instruction (ret.) will be omitted hereafter)

5 0= . 0 (equal to zero)
0 0= . 1 (not equal to zero)

Example 10-7

10 > IF ." BIGGER " THEN ." SMALLER "

If the top of the stack is greater than 10 when this statement is encountered the flag left by > is TRUE, hence BIGGER is printed. Otherwise the word SMALLER is printed but not BIGGER.

Note, however, that the IF-THEN can only be used inside a *colon definition*, a concept we will consider in the next section. Example 10-6 will not work if you type it into a terminal unless it is within the definition of a new FORTH word.

You should note that the above example also has the result that a TRUE flag prints both BIGGER and SMALLER while a FALSE flag only prints SMALLER. The words after THEN are always executed, no matter what the flag is. To allow a completely clean choice of branches, the -ELSE construct is used:

...flag..IF..(execute TRUE)..ELSE..(execute FALSE)..THEN...

Example 10-8

...10 > IF ." BIGGER " ELSE ." SMALLER " THEN ....

will result in BIGGER or SMALLER, but never both, being printed depending on whether the stack top was larger than 10 or smaller.

FORTH also permits looping. An *indefinite loop* is one that is again conditional on a flag. The BEGIN-UNTIL structure is:

...BEGIN...(execute words)..flag..UNTIL...

Example 10-9

... + 0 BEGIN CR DUP . 1+ DUP 10 = UNTIL ." DONE"...

This will start by printing 0 and increase the printout by 1 until the sum reaches 10. (Note that the single FORTH word (1+) increments the top of the stack by 1 in the same way as would (1) (+). Until the sum reaches 10, the program encounters a FALSE flag (0) which causes it to loop back to the CR after

BEGIN. When 10 is reached, the flag becomes TRUE and halts the looping to execute the statement after UNTIL, printing DONE.

Another form of indefinite loop is BEGIN-WHILE-REPEAT. This structure executes code between WHILE and REPEAT as long as the flag just before WHILE is TRUE. It then branches unconditionally back to BEGIN and repeats the test. If the flag becomes FALSE, it skips to just after REPEAT.

Example 10-10

```
0 BEGIN CR DUP . DUP 10 = NOT WHILE 1+ ." MORE "
REPEAT ." DONE "
```

will print 0 to 9 followed by MORE and on the tenth cycle, 10 DONE.

Again, none of the conditional words BEGIN UNTIL WHILE REPEAT are effective except inside a colon definition.

10-8. FORTH Finite Loops

The finite loop as in BASIC will repeat a portion of code for a specific number of times. There are several variants, the simplest being:

```
...limit initial DO...(execute words)...LOOP...
```

Starting the index with the initial value on top of the stack, the words between DO and LOOP are executed and an index is increased by 1 repeatedly until the value of limit is reached. Control then passes to the statement following LOOP. The current value of the index is held in a variable I which can be placed on the stack and printed by the word I.

Example 10-11

```
...10 0 DO I . LOOP...
```

will print the numbers 0 to 9 when executed.

```
0 1 2 3 4 5 6 7 8 9
```

Note that the limit value placed on the stack (10 in this case) is one more than the count achieved. This is because the index is incremented at the word LOOP and control passed beyond it when the index reaches limit.

It is possible in FORTH to put a DO loop within a DO loop for nesting up to three deep. The index variables are I,K,J. One may also increment a DO loop by an integer other than 1, or even decrement (count downward) with a negative integer. This is accomplished by using the construction n +LOOP rather than LOOP, n being the new incremental step.

10-9. Colon Definitions

The most important and novel feature of FORTH is the ability to create new definitions. The *colon definition* form is one way of accomplishing this. It consists of a colon, (:), name, definition, and a terminating semicolon (;).

: name ...(definition) ... ;

Example 10-12

The DO loop shown previously can be given a name COUNT and defined as:

```
: COUNT 0 DO I . LOOP ;
```

Typing into the terminal

```
10 COUNT
```

will result in

```
0 1 2 3 4 5 6 7 8 9
```

as before. COUNT will count to one less than whatever number precedes it on the stack, e.g.

5 COUNT (ret.) <u>0 1 2 3 4</u>

The word EXIT terminates the execution of a colon definition and permits one to leave an IF statement, for example. But EXIT is invalid within a DO loop, which is terminated by the word LEAVE.

10-10. List Search Program in FORTH

FORTH commands and definitions entered into the terminal as described above cannot be edited or "saved" and so will be lost as soon as power is removed from the computer.

To edit and store FORTH programs, the statements are organized into *screens*. Each screen is a block of 1024 characters subdivided into 16 lines of 64 characters each, numbered 0 to 15. Each screen is given a number for identification and retrieval. A program may consist of one or more screens. FORTH systems implemented in various types of computers have different editors, which will not be described here. The editor permits the screens to be written, formatted, corrected, and supplied with comments enclosed in parentheses ( ). Although FORTH is a free-form language like BASIC, the use of proper screen editing conventions results in a "structured" program appearance that is easier to read.

```
Screen 3        HEX 3

 0   (LIST SEARCH DEFINITIONS—1                    RJB 5/16/82)
 1   FORTH DEFINITIONS
 2
 3   (REQUESTS INPUT OF LIST FROM TERMINAL AND              )
 4   (ENTRY OF KEY. SEARCHES LIST FOR KEY                   )
 5   (PRINTING OUT POSITION OR IF KEY NOT FOUND             )
 6
 7   : INPUT QUERY INTERPRET ; (ACCEPTS TERMINAL INPUT )
 8   : INLIST ." ENTER LIST TO SEARCH "
            INPUT S. ;              (ENTER AND VERIFY LIST )
10   VARIABLE N                    (DEFINES ADDRESS ON N    )
11   : SETUP DEPTH 1 + N ! ; (SETS UP SEARCH LIMIT          )
12   VARIABLE KY                   (DEFINES ADDRESS OF KY  )
13   : KEYIN CR ." ENTER SEARCH KEY "
14         INPUT KY ! ;            (STORES KEY IN VARIABLE )
15   -->
```

FIG. 10-1. First FORTH Screen

Example 10-13

Figure 10-1 is a picture of the first screen of a program to search lists and is almost identical to the BASIC program studied in the previous unit. Two screens are used for the program, but it could have been compressed into one. It is not considered good practice to crowd screens, so space is left for blank lines (and possible expansion of the program) and for comments placed within the round parentheses.

Line zero is a comment-style title that describes what is on the screen. A FORTH word INDEX lists the first line of all screens in memory, so this is an important line. The second line states the name of the vocabulary, FORTH, which specifies the lists of definitions to be placed in the dictionary and searched when compiling new definitions. Other vocabularies may exist, such as EDITOR, used when editing.

Lines three to five are comments and lines two and six are left blank for ease of reading.

The first definition is on line seven. The predefined dictionary word QUERY allows up to 80 characters to be input from the terminal, while INTERPRET allows these characters to be compiled; if numbers, they are placed on the stack. Thus, the word INPUT allows a list of numbers to be placed on the stack. This will be the list to be searched by the program and has the same meaning as the DATA list (line ten of the BASIC program) that we saw in the last unit. The terminal input is an improvement over the BASIC program.

Line eight defines the word INLIST. This merely prompts the user to input the list by printing the message.

ENTER LIST TO SEARCH

and then uses INPUT to accept what follows from the terminal. Using the newly defined word INPUT within the body of the newer definition INLIST is typical of FORTH's "threaded" nature.

VARIABLE N is next defined on line nine but given no value as yet. On line eleven we evaluate N, which will be the upper limit of a DO loop in the next screen. The DO loop is going to be used to search every number on the list, but we also want to terminate the search if it reaches the last item without finding the key. (See Knuth's algorithm in Unit 9.) Since we don't know in advance how many items will enter the list, we use the predefined word DEPTH to find out how many are on the stack. (It is assumed the stack is empty before running this program. This can be assured by first typing the word ABORT.)

DEPTH places on top of the stack the number of items that were in it, but to this we must add 1 to get the proper value of limit for a DO loop to the bottom of the stack. DEPTH 1 + computes the correct value for N. This is stored in N by the "store" word (!). This completes the definition of SETUP.

VARIABLE KY is defined to allow the user to enter the value of the key to be searched for. KEYIN is of the same form as INLIST, printing the user prompt and following it INPUT. The number put on the stack by INPUT is stored in KY, using (!).

```
Screen 4     HEX 4
  0          (LIST SEARCH DEFINITIONS—2              RJB 5/16/82)
  1
  2          (THIS SCREEN PERFORMS ACTUAL SEARCH USING )
  3          (DO LOOP THROUGH STACK. IF KEY EQUALS      )
  4          (CURRENT TOP OF STACK, WE LEAVE THE LOOP.  )
  5
  6        : LOOK N @ 1 DO                (FROM, 1 TO N LOOP)
  7                  KY @ =              (IF KEY = STACK TOP)
  8                  IF CR ." KEY IS "   (PRINT THE KEY'S     )
  9                    I . LEAVE         (INDEX & LEAVE LOOP.)
 10                  ELSE N @ 1 - I =   (IF LAST ON STACK    )
 11                    IF CR ." KEY NOT FOUND "
 12                    THEN                 (END OF IF — THEN )
 13                  THEN (END OF IF -ELSE - THEN              )
 14                         LOOP ; (REPEAT DO LOOP OR END )
 15        : SEARCH INLIST SETUP KEYIN LOOK ; (THAT'S ALL!)
```

FIG. 10-2. Second FORTH Screen

This screen then takes care of all the preliminaries. The variables are now set up to control a DO loop through the stack. This loop is pictured on Screen 4, Fig. 10-2. (The arrow symbol ——> on line 15 of Screen 3 means to continue compiling with the next screen.)

The search is performed by the word LOOK. This is defined as a DO loop from 1 to (N−1), the stack depth. At each step the value stored in KY is fetched and the top of the stack removed and compared with it. Control then passes to an IF-ELSE-THEN conditional. If the comparison is TRUE, lines eight and nine print out the value of I, which is the location (measuring from the top of the stack) where the key was found. The word LEAVE permits the DO loop to be terminated by setting the value of I to N. This terminates the program.

If the comparison is FALSE (the key was not found at this step), another nested IF statement is used to find if we have yet reached the bottom of the stack. If so, the value of I will have reached N-1 (or DEPTH). As the statement is TRUE, the program prints out KEY NOT FOUND. But if it is not TRUE, we still have more of the list to search, so after terminating the two IF clauses with THENs, we reach LOOP and return to the statement after DO. Since we discarded the top of the stack when we performed the (=) test, we now have a new number on top to test.

The FORTH program is "wrapped up" in a single word on line 15. SEARCH combines all the operations of INLIST, SETUP, and LOOP in that sequence. To run the program, then, we merely type in:

SEARCH

Exercises

10-1. *Review Example 10-4. Now write the FORTH expression for the circumference (2*PI*R) and give the result.*

10-2. *What is in the stack after entering each of the following groups? (Assume the stack is empty before each group enters.)*

1 3 5 DUP

1 3 5 DUP DROP

1 3 5 ROT DUP

1 3 5 SWAP .

1 3 5 DUP .

10-3. *What is the terminal output after each of the following?*

5 10 5 + .

1 3 5 * .

6 8 2 / .

6 2 8 / .

5 3 1 + * .

5 7 3 / .

4 6 8 * SWAP / .

10-4. *a) Declare a constant 12 called DOZEN. Using FORTH, multiply it by 3 and print out 36.*
b) Multiply a constant WIDTH equal to 5 by three values of DEPTH: 5,10,15; and print out the results.

10-5. *In Example 10-3 the single FORTH word (*/) is used to obtain the 16-bit result 78. If two separate words, (*) followed by (/) are substituted is the result the same? Why?*

10-6. *Write a colon definition that will print the numbers 1 through 9, using a BEGIN-WHILE-REPEAT construction.*

10-7 *Write a colon definition that will test the top of the stack and print GREATER, EQUAL, or SMALLER, depending on whether the stack top is greater, equal, or smaller than 5. Use nested IF type conditional statements. (Hint: Use an EXIT to prevent double labeling of results.)*

10-8. *What would be the result if 1 were not added to DEPTH in the example 10-11, Screen 3, line 11 (see Fig. 10-1).*

References

Katzan, Harry Jr., *Invitation to FORTH*, New York: Petrocelli (1981).
Brodie, Leo, *Starting FORTH*, Mt. View, California: Mountain View Press (1981).
Brown, Jack W., *SK-FORTH User's Guide*, New Westminister, B.C., Canada: Saturn Software, Ltd. (1982).

Unit 11: Digital Control Algorithms

UNIT 11

Digital-Control Algorithms

The purpose of this unit is to show how to develop digital-control algorithms that can be programmed in high-level languages such as BASIC or FORTH for use with microprocessor systems. It is assumed that the reader is familiar with process control and controller algorithms as explained in Unit 5 of the ISA Independent Learning Module *Fundamentals of Process Control Theory* by Dr. Paul W. Murrill (Ref. 1) and particularly the PID (ideal) algorithm $K(1+1/T_iS+T_dS)$.

Learning Objectives — When you have completed this unit you should:

A. Be able to convert dynamic analog control algorithms to difference equations for digital processing.

B. Know how control equations are programmed in a high-level language.

C. Understand a digital-processor program for the PID algorithm.

11-1. Difference Equations

A digital controller differs from its analog prototype in that the digital instrument samples the process variable and outputs a calculated valve-control signal at discrete intervals rather than continuously. The analog controller is governed by a continuous algebraic equation containing integral and/or derivative terms which describe the dynamic (time-varying) behavior of the input or error signal. An equivalent method of describing dynamic components is the block diagram (Ref. 2) which is a graphic picture of the *transfer function*: the response (ratio) of the output of the block or box to its input. Figure 11-1 is an example of a block diagram describing an RC electrical filter, a simple first-order time lag. (This example and the following is taken from the author's book *Microprocessors in Instruments and Control* (Ref. 3) to which the reader is referred for a fuller discussion.)

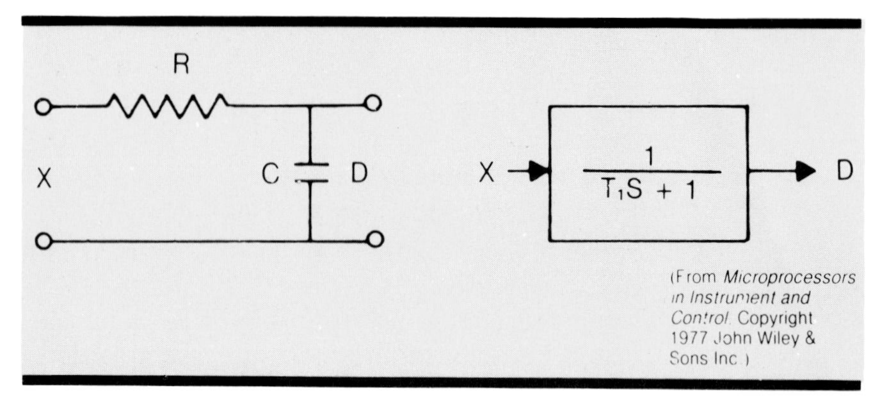

Fig. 11-1. Single-Lag Filter

The effect of a sudden rise (step) of the voltage X across the input terminals is an exponential increase in the output D, with a slope determined by the time constant RC (called T[1] in the block diagram, see Fig. 11-2). S is the so-called Laplace variable, indicating the mathematical operation of differentiation with respect to time, so that the expression SD is a shorthand way of expressing the derivative dD/dt. (Instead of S, one can use p, the Heaviside operator, or d/dt, as in Ref. 1.)

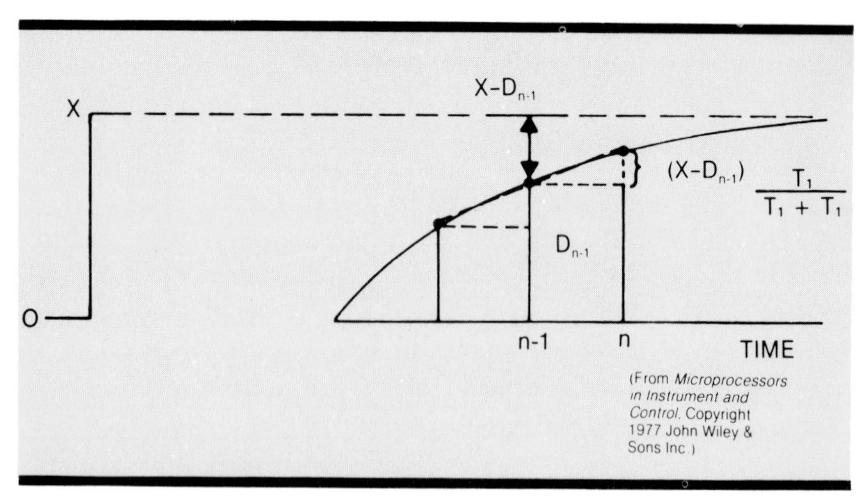

Fig. 11-2. Digital Filter Output

When developing an equivalent control equation for the digital microprocessor we must find a method of transforming the continuous differential equation (containing the derivative) into one that changes with time in discrete jumps. Such an expression is called a *difference equation*. The interval between jumps in the input and output is called the *sampling time*. Clearly, the smaller the sampling interval and the more jumps per second, the closer we will come to the continuous dynamic equation of the original. But for each sampling interval there must be a calculation by the microprocessor of what the next output should be. There are limits to the computing speed of even a fast machine and efficient program, especially if several tasks or control loops are to be computed in the same time interval.

As discussed briefly in Unit 2, if there are lower limits to sampling time, "common sense" would tell us that digital control can only approach the performance of its analog counterpart and so can never be as "good." Common sense would be wrong in this instance. A fundamental principle of *information theory*, the underlying basis for much of computing and control theory, is known as Shannon's Theorem, after Claude Shannon, a Bell Telephone Laboratories scientist who was one of the founders of this discipline. Shannon's Theorem states that a signal that is bandwidth-limited, containing no frequency components greater than w, can be recovered without any distortion or loss from samples taken at a rate twice that of highest frequency ($2w$). Since real plant processes are definitely limited in bandwidth (for example, some distillation processes have time constants of over an hour), this means that a sampling time equivalent to twice the highest plant frequency is adequate. In practice, sampling rates of 0.3 to 0.5 sec prove fast enough for the vast majority of process control problems experienced.

Example 11-1

From block diagrams similar to those in Ref. 1 we can write the transfer function directly. For example, the transfer function of Fig. 11-1 is the ratio of the box output D to its input X, or

$$D/X = 1/(T[1]S+1)$$

Consequently

$$T[1]S*D+D = X \qquad (11\text{-}1)$$

Because S means the operation d/dt we can also write

$$T[1]dD/dt+D = X \qquad (11\text{-}2)$$

The last is a differential equation. To obtain the *difference equation* which is suitable for a digital algorithm, we substitute for the derivative dD/dt the approximation

$$(D[n]-D[n-1])/T[s]$$

where D[n] means the value of the output D at the present time interval [n] and [n−1] refers to the previous time interval, etc. T[s] is the time between sampling instances (sampling interval) in seconds, and is equal to

$$T[n]-T[n-1]$$

for any n. T[s] is assumed to remain constant for a given task.

Using these approximations we can transform Eq. (11-2) into a difference equation:

$$(T[1]\,(D[n]-D[n-1])\,/\,T[s]+D[n] = X \qquad (11\text{-}3)$$

With some algebraic manipulation (see Ref. 3), we can solve this equation explicity for the current value of output D[n] as a function of current input and the past values of output and input. This is a possible or *realizable* solution to the equation, because naturally any expression for D[n] that required future values of input or output would require a "crystal ball" and be impossible to use.

The useful or realizable solution of Eq. (11-3) is, therefore,

$$D[n] = D[n-1]+(T[s]/(T[s]+T[1]))*(X-D[n-1]) \qquad (11\text{-}4)$$

This is the equation that the digital processor must solve repeatedly, once for each sampling interval, in order to keep the output D[n] up to date for increasing n's. The inputs to this operation are the current value of X (the initial voltage rise step value), the value of the last output D[n−1], the sampling interval T[s], and the filter time constant or RC value, T[1]. By solving this equation repetitively, the processor simulates the behavior of the analog RC filter shown in Fig. 11-1. The output will change with time in a step-wise fashion for each discrete interval of time

$$\ldots\ [n-2],\ [n-1],\ [n].$$

As seen in Fig. 11-2, D[ ] approaches the value X asymptotically. From Eq. 11-4 we see that each successive increase in D builds up on the previous value D[n−1] by an amount equal to the remaining different (X−D[n−1]) multiplied by the "filter constant." This is equal to the ratio T[s]/(T[s]+T[1]). The rate at which the output approaches the step change X is thus governed by the RC time constant T[1], just as in the analog filter. As T[1] is made smaller, the rate (slope) of the filter voltage output becomes greater.

11-2. Digital Algorithms, General Method

The method used to develop any digital algorithm from a continuous equation containing differential and integral terms can be generalized in the following steps.

1. The integrodifferential equation of the analog form of the algorithm is obtained from the block diagram transfer function or the Laplace transform (see Ref. 1 for examples).

2. The analog equation is put into difference form, using the approximations

   Derivative: dD/dt = (D[n]−D[n−1])/T[s]

   Integral: integral $\int$ (edt) = summation $\sum$ e[i]T[s]

3. Solve for current value of output using only the current or past values of input and error and past values of output in the right-hand side.

Example 11-2

The transfer function of the ideal (noninteracting) form of the PID equation is shown in Fig. 11-3. (See Ref. 1, p. 62)

V[n]/E[n] = K+KI/S+KDS (11-5)

where V is the valve position (output) and E is the error (PV-SP), while K is the proportional gain. But KI, the integral gain constant, equals K/TI and KD, the derivative gain is K*TD (see Ref. 1 or Ref. 4, p. 160 if this is not clear). TI and TD are the integral and derivative time constant adjustments of the controller. So, applying Steps 1 and 2 above

$$V[n]=K(E[n]+(1/TI)*\sum E[i]T[s]+TD(E[n]-E[n-1])/T[s]) \qquad (11\text{-}6)$$

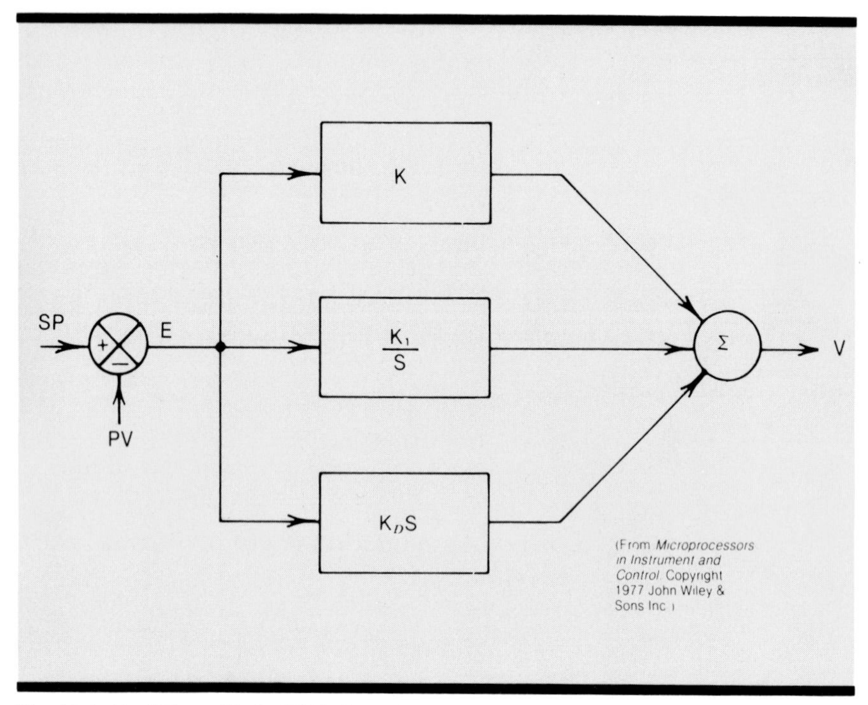

Fig. 11-3. Ideal Three-Mode (PID) Controller

11-3. Program for Ideal PID Controller

The PID digital algorithm can be programmed directly in a high-level language suitable for running on a microprocessor. The program listed below was originally written in FORTRAN but is almost identical when changed to BASIC except for the variable names. In both languages it is written as a subroutine so that the inputs (setpoint SP and process variable PV) can be measured and fed back to the subroutine, repetitively calculating the new valve-position output. This subroutine is used in a closed-loop simulator to test the behavior of various control algorithms and processes. With little modification, it can be used to program a multiple-loop controller.

```
1000  REM IDEAL PID CONTROLLER
1010  REM
1020  REM PRIOR TO ENTERING THIS SUBPROGRAM THE FOLLOWING
1030  REM CONSTANT VALUES MUST BE INPUT:
1040  REM K = CONTROLLER GAIN
1050  REM TI = INTEGRAL TIME
1060  REM TD = DERIVATIVE TIME
1070  REM TS = SAMPLE INTERVAL TIME
1080  REM TF = FILTER TIME CONSTANT (E.G. 0.01)
1090  REM
3000  REM INITIALIZE
3010  REM VALUES FOR PV AND SP ARE ASSUMED AVAILABLE FROM
3020  REM A DATA ACQUISITION SUBPROGRAM
3030  REM
3040  F1 = PV - SP
3050  F0 = F1
3060  REM
3070  REM PUT SUBPROGRAM FOR DATA ACQUISITION HERE
3080  REM
5000  REM PID CONTROL SUBROUTINE
5010  REM THIS PROGRAM IS CALLED BY THE MAIN PROGRAM ON A
5020  REM MULTIPLE OF THE SAMPLING INTERVAL TS
5030  REM
5040  REM
5050  E2 = PV - SP
5060  F2 = (TF*F1 - TS*E2)/(TF + TS)
5070  V = K*(F2 - F1) + F2*TS + TD*(F2 - F1 - F1 + F0)/TS)
5080  F0 = F1
5090  F1 = F2
5100  V2 = V2 - V
5110  REM V2 IS UPDATED VALVE POSITION
5120  RETURN
5130  REM RETURNS V2 AND CONTROL TO MAIN PROGRAM
```

Table 11-1. Digital PID Algorithm

Study of this program will reveal the general method of programming control algorithms for digital computers. It is left as an (unofficial) exercise for the reader to rewrite this program in FORTH and compare it with the BASIC version.

With relatively minor alterations, the algorithm could be changed to the "real" or interactive PID form. Dynamic compensation (lead/lag) for feedforward loops, dead-time compensation, or even optimal algorithms such as the minimal prototype Kalman filter or "dead beat" can be programmed just as readily. All of these algorithms have been worked out and are given in digital (difference equation) form in Ref. 5, but are too lengthy to include here.

It also should be noted that there are more sophisticated and powerful mathematical methods of deriving digital algorithms, especially useful for advanced types. The chief method used is the Z-transform, a digital version of the Laplace transform method, but this assumed a mathematical skill beyond that assumed in this ILM.

In Closing

Congratulations! You have made it through the ILM *Microprocessors in Industrial Control*. Although this does not yet qualify you as an expert in computer design or programming, it should give you a sufficient overview of the subject to be able to understand microprocessor applications and to learn enough to undertake your own plant projects from the plentiful literature on microprocessors and programming.

Exercises

11-1. *High-fidelity audio recordings are now made using digital methods, by sampling the output of a microphone at discrete intervals. Given a sampling time greater than zero, can they be as good as analog recordings? Better? Suggest a sampling frequency and give reasons.*

11-2. *Discuss the effect of varying sampling time* $T[s]$ *on the digital filter of Fig. 11-1 and Eq. (11-4).*

11-3. *In Eq. (11-4), Example 11-1, we derived the digital equivalent of a simple RC filter. Can you think of any reasons why we would want to use a microprocessor to perform such a simple function?*

11-4. *In Example 11-2, why was Step 3 omitted?*

11-5. *What line of the BASIC program in Table 11-1 is the equivalent of Eq. 11-6?*

11-6. *What is the function of line 5060 in the above program? (Hint: compare with Eq. (11-4).)*

11-7. *What expressions in line 5070 of the BASIC program given are used as the error inputs* $E[i]$ *of Eq. (11-6)?*

11-8. *What is the function of lines 5080 through 5090?*

References

[1]Murrill, P. W., *Fundamentals of Process Control Theory*, Research Triangle Park, North Carolina: Instrument Society of America (1981), pp. 52-62
[2]Ibid., pp. 23-28.
[3]Bibbero, R. J., *Microprocessors in Instruments and Control*, New York: John Wiley & Sons, (1977), pp. 99-100.
[4]Ibid., Chapter 6, "Development of Digital Control Algorithms."
[5]Ibid., pp. 162-175.

Appendix A: Suggested Readings and Study Materials

APPENDIX A

Suggested Readings and Study Materials

Independent Learning Modules

Murrill, P. W. *Fundamentals of Process Control Theory* (Instrument Society of America, 1981).

Handbooks

Lien, D. L. *The BASIC Handbook* (Compusoft, 1978).

Textbooks

Bibbero, R. J. *Microprocessors in Instruments and Control* (Wiley-Interscience, 1977).

Bibbero, R. J. and Stern, D. M. *Microprocessor Systems, Interfacing and Applications* (Wiley-Interscience, 1982).

Brodie, L. *Starting FORTH* (Mountain View Press, 1981).

Leventhal, L. A. *6502 Assembly Language Programming* (Osborne-McGraw-Hill, 1979).

Osborne, A. *An Introduction to Microcomputers, Vol. 1, Basic Concepts* (Osborne, 1976).

Katzan, H. *Introduction to FORTH* (Mountain View Press, 1981).

Skier, K. *Beyond Games, System Software for Your 6502 Personal Computer* (BYTE/McGraw-Hill, 1981).

Appendix B: Instruction List Alphabetic by Mnemonic with Op Codes, Execution Cycles and Memory Requirements

The following notation applies to this summary:

A	Accumulator
X, Y	Index Registers
M	Memory
P	Processor Status Register
S	Stack Pointer
√	Change
_	No Change
+	Add
∧	Logical AND
-	Subtract
⊻	Logical Exclusive Or
↑	Transfer from Stack
↓	Transfer to Stack
→	Transfer to
←	Transfer to
V	Logical OR
PC	Program Counter
PCH	Program Counter High
PCL	Program Counter Low
OPER	OPERAND
#	IMMEDIATE ADDRESSING MODE

Note: At the top of each table is located in parentheses a reference number (Ref: XX) which directs the user to that Section in the MCS6500 Microcomputer Family Programming Manual in which the instruction is defined and discussed.

ADC *Add memory to accumulator with carry* ADC

Operation: A + M + C → A, C

N Z C I D V

✓ ✓ ✓ — — ✓

(Ref: 2.2.1)

Addressing Mode	Assembly Language Form	OP CODE	No. Bytes	No. Cycles
Immediate	ADC # Oper	69	2	2
Zero Page	ADC Oper	65	2	3
Zero Page, X	ADC Oper, X	75	2	4
Absolute	ADC Oper	6D	3	4
Absolute, X	ADC Oper, X	7D	3	4*
Absolute, Y	ADC Oper, Y	79	3	4*
(Indirect, X)	ADC (Oper, X)	61	2	6
(Indirect), Y	ADC (Oper), Y	71	2	5*

* Add 1 if page boundary is crossed.

AND *"AND" memory with accumulator* AND

Logical AND to the accumulator

Operation: A ∧ M → A

N Z C I D V

✓ ✓ — — — —

(Ref: 2.2.3.0)

Addressing Mode	Assembly Language Form	OP CODE	No. Bytes	No. Cycles
Immediate	AND # Oper	29	2	2
Zero Page	AND Oper	25	2	3
Zero Page, X	AND Oper, X	35	2	4
Absolute	AND Oper	2D	3	4
Absolute, X	AND Oper, X	3D	3	4*
Absolute, Y	AND Oper, Y	39	3	4*
(Indirect, X)	AND (Oper, X)	21	2	6
(Indirect), Y	AND (Oper), Y	31	2	5

* Add 1 if page boundary is crossed.

ASL

ASL *Shift Left One Bit (Memory or Accumulator)*

ASL

Operation: C ← | 7 | 6 | 5 | 4 | 3 | 2 | 1 | Ø | ← Ø

N Z C I D V
✓ ✓ ✓ – – –

(Ref: 10.2)

Addressing Mode	Assembly Language Form	OP CODE	No. Bytes	No. Cycles
Accumulator	ASL A	ØA	1	2
Zero Page	ASL Oper	Ø6	2	5
Zero Page, X	ASL Oper, X	16	2	6
Absolute	ASL Oper	ØE	3	6
Absolute, X	ASL Oper, X	1E	3	7

BCC

BCC *Branch on Carry Clear*

BCC

Operation: Branch on C = Ø

N Z C I D V
– – – – – –

(Ref: 4.1.1.3)

Addressing Mode	Assembly Language Form	OP CODE	No. Bytes	No. Cycles
Relative	BCC Oper	9Ø	2	2*

* Add 1 if branch occurs to same page.

* Add 2 if branch occurs to different page.

BCS

BCS *Branch on carry set*

Operation: Branch on C = 1

N Z C I D V
_ _ _ _ _ _

(Ref: 4.1.1.4)

Addressing Mode	Assembly Language Form	OP CODE	No. Bytes	No. Cycles
Relative	BCS Oper	BØ	2	2*

* Add 1 if branch occurs to same page.

* Add 2 if branch occurs to next page.

BEQ

BEQ *Branch on result zero*

Operation: Branch on Z = 1

N Z C I D V
_ _ _ _ _ _

(Ref: 4.1.1.5)

Addressing Mode	Assembly Language Form	OP CODE	No. Bytes	No. Cycles
Relative	BEQ Oper	FØ	2	2*

* Add 1 if branch occurs to same page.

* Add 2 if branch occurs to next page.

BIT

BIT *Test bits in memory with accumulator*

Operation: $A \wedge M$, $M_7 \rightarrow N$, $M_6 \rightarrow V$

Bit 6 and 7 are transferred to the status register. If the result of $A \wedge M$ is zero then Z = 1, otherwise Z = Ø

N Z C I D V
M_7 ✓ — — — M_6

(Ref: 4.2.1.1)

Addressing Mode	Assembly Language Form	OP CODE	No. Bytes	No. Cycles
Zero Page	BIT Oper	24	2	3
Absolute	BIT Oper	2C	3	4

BMI

BMI *Branch on result minus*

Operation: Branch on N = 1

N Z C I D V
\- - - - - -

(Ref: 4.1.1.1)

Addressing Mode	Assembly Language Form	OP CODE	No. Bytes	No. Cycles
Relative	BMI Oper	3Ø	2	2*

Add 1 if branch occurs to same page.

* Add 2 if branch occurs to different page.

BNE

BNE *Branch on result not zero*

Operation: Branch on Z = 0

N Z C I D V
\- - - - - -

(Ref: 4.1.1.6)

Addressing Mode	Assembly Language Form	OP CODE	No. Bytes	No. Cycles
Relative	BNE Oper	DØ	2	2*

* Add 1 if branch occurs to same page.

* Add 2 if branch occurs to different page.

BPL

BPL *Branch on result plus*

Operation: Branch on N = Ø

N Z C I D V
\- - - - - -

(Ref: 4.1.1.2)

Addressing Mode	Assembly Language Form	OP CODE	No. Bytes	No. Cycles
Relative	BPL Oper	1Ø	2	2*

* Add 1 if branch occurs to same page.

* Add 2 if branch occurs to different page.

BRK

BRK *Force Break*

Operation: Forced Interrupt PC + 2 ↓ P ↓

N Z C I D V
— — — 1 — —

(Ref: 9.11)

Addressing Mode	Assembly Language Form	OP CODE	No. Bytes	No. Cycles
Implied	BRK	00	1	7

1. A BRK command cannot be masked by setting I.

BVC

BVC *Branch on overflow clear*

Operation: Branch on V = 0

N Z C I D V
— — — — — —

(Ref: 4.1.1.8)

Addressing Mode	Assembly Language Form	OP CODE	No. Bytes	No. Cycles
Relative	BVC Oper	50	2	2*

* Add 1 if branch occurs to same page.
* Add 2 if branch occurs to different page.

BVS

BVS *Branch on overflow set*

Operation: Branch on V = 1

N Z C I D V
— — — — — —

(Ref: 4.1.1.7)

Addressing Mode	Assembly Language Form	OP CODE	No. Bytes	No. Cycles
Relative	BVS Oper	70	2	2*

* Add 1 if branch occurs to same page.
* Add 2 if branch occurs to different page.

CLC — CLC *Clear carry flag* — CLC

Operation: Ø → C

(Ref: 3.0.2)

N Ƶ C I D V
— — Ø — — —

Addressing Mode	Assembly Language Form	OP CODE	No. Bytes	No. Cycles
Implied	CLC	18	1	2

CLD — CLD *Clear decimal mode* — CLD

Operation: Ø → D

(Ref: 3.3.2)

N Ƶ C I D V
— — — — Ø —

Addressing Mode	Assembly Language Form	OP CODE	No. Bytes	No. Cycles
Implied	CLD	D8	1	2

CLI — CLI *Clear interrupt disable bit* — CLI

Operation: Ø → I

(Ref: 3.2.2)

N Ƶ C I D V
— — — Ø — —

Addressing Mode	Assembly Language Form	OP CODE	No. Bytes	No. Cycles
Implied	CLI	58	1	2

CLV

CLV *Clear overflow flag*

Operation: Ø → V

N Z C I D V

– – – – – Ø

(Ref: 3.6.1)

Addressing Mode	Assembly Language Form	OP CODE	No. Bytes	No. Cycles
Implied	CLV	B8	1	2

CMP

CMP *Compare memory and accumulator*

Operation: A - M

N Z C I D V

✓ ✓ ✓ – – –

(Ref: 4.2.1)

Addressing Mode	Assembly Language Form	OP CODE	No. Bytes	No. Cycles
Immediate	CMP #Oper	C9	2	2
Zero Page	CMP Oper	C5	2	3
Zero Page, X	CMP Oper, X	D5	2	4
Absolute	CMP Oper	CD	3	4
Absolute, X	CMP Oper, X	DD	3	4*
Absolute, Y	CMP Oper, Y	D9	3	4*
(Indirect, X)	CMP (Oper, X)	C1	2	6
(Indirect), Y	CMP (Oper), Y	D1	2	5*

* Add 1 if page boundary is crossed.

CPX

CPX *Compare Memory and Index X*

CPX

Operation: X - M

N Z C I D V
✓ ✓ ✓ - - -

(Ref: 7.8)

Addressing Mode	Assembly Language Form	OP CODE	No. Bytes	No. Cycles
Immediate	CPX #Oper	EØ	2	2
Zero Page	CPX Oper	E4	2	3
Absolute	CPX Oper	EC	3	4

CPY

CPY *Compare memory and index Y*

CPY

Operation: Y - M

N Z C I D V
✓ ✓ ✓ - - -

(Ref: 7.9)

Addressing Mode	Assembly Language Form	OP CODE	No. Bytes	No. Cycles
Immediate	CPY #Oper	CØ	2	2
Zero Page	CPY Oper	C4	2	3
Absolute	CPY Oper	CC	3	4

DEC

DEC *Decrement memory by one*

DEC

Operation: M - 1 → M

N Z C I D V
✓ ✓ - - - -

(Ref: 10.7)

Addressing Mode	Assembly Language Form	OP CODE	No. Bytes	No. Cycles
Zero Page	DEC Oper	C6	2	5
Zero Page, X	DEC Oper, X	D6	2	6
Absolute	DEC Oper	CE	3	6
Absolute, X	DEC Oper, X	DE	3	7

DEX

DEX *Decrement index X by one*

Operation: X - 1 → X

N Z C I D V
✓ ✓ - - - -

(Ref: 7.6)

Addressing Mode	Assembly Language Form	OP CODE	No. Bytes	No. Cycles
Implied	DEX	CA	1	2

DEY

DEY *Decrement index Y by one*

Operation: Y - 1 → Y

N Z C I D V
✓ ✓ - - - -

(Ref: 7.7)

Addressing Mode	Assembly Language Form	OP CODE	No. Bytes	No. Cycles
Implied	DEY	88	1	2

EOR

EOR *"Exclusive-Or" memory with accumulator*

Operation: A ⊻ M → A

N Z C I D V
✓ ✓ - - - -

(Ref: 2.2.3.2)

Addressing Mode	Assembly Language Form	OP CODE	No. Bytes	No. Cycles
Immediate	EOR #Oper	49	2	2
Zero Page	EOR Oper	45	2	3
Zero Page, X	EOR Oper, X	55	2	4
Absolute	EOR Oper	4D	3	4
Absolute, X	EOR Oper, X	5D	3	4*
Absolute, Y	EOR Oper, Y	59	3	4*
(Indirect, X)	EOR (Oper, X)	41	2	6
(Indirect),Y	EOR (Oper), Y	51	2	5*

* Add 1 if page boundary is crossed.

INC

INC *Increment memory by one*

Operation: M + 1 → M

N Z C I D V
✓ ✓ – – – –

(Ref: 10.6)

Addressing Mode	Assembly Language Form	OP CODE	No. Bytes	No. Cycles
Zero Page	INC Oper	E6	2	5
Zero Page, X	INC Oper, X	F6	2	6
Absolute	INC Oper	EE	3	6
Absolute, X	INC Oper, X	FE	3	7

INX

INX *Increment Index X by one*

Operation: X + 1 → X

N Z C I D V
✓ ✓ – – – –

(Ref: 7.4)

Addressing Mode	Assembly Language Form	OP CODE	No. Bytes	No. Cycles
Implied	INX	E8	1	2

INY

INY *Increment Index Y by one*

Operation: Y + 1 → Y

N Z C I D V
✓ ✓ – – – –

(Ref: 7.5)

Addressing Mode	Assembly Language Form	OP CODE	No. Bytes	No. Cycles
Implied	INY	C8	1	2

JMP

JMP *Jump to new location* **JMP**

Operation: (PC + 1) → PCL
(PC + 2) → PCH

(Ref: 4.0.2)
(Ref: 9.8.1)

N Z C I D V
— — — — — —

Addressing Mode	Assembly Language Form	OP CODE	No. Bytes	No. Cycles
Absolute	JMP Oper	4C	3	3
Indirect	JMP (Oper)	6C	3	5

JSR

JSR *Jump to new location saving return address* **JSR**

Operation: PC + 2 ↓, (PC + 1) → PCL
(PC + 2) → PCH

(Ref: 8.1)

N Z C I D V
— — — — — —

Addressing Mode	Assembly Language Form	OP CODE	No. Bytes	No. Cycles
Absolute	JSR Oper	20	3	6

LDA

LDA *Load accumulator with memory* **LDA**

Operation: M → A

N Z C I D V
✓ ✓ — — — —

(Ref: 2.1.1)

Addressing Mode	Assembly Language Form	OP CODE	No. Bytes	No. Cycles
Immediate	LDA # Oper	A9	2	2
Zero Page	LDA Oper	A5	2	3
Zero Page, X	LDA Oper, X	B5	2	4
Absolute	LDA Oper	AD	3	4
Absolute, X	LDA Oper, X	BD	3	4*
Absolute, Y	LDA Oper, Y	B9	3	4*
(Indirect, X)	LDA (Oper, X)	A1	2	6
(Indirect), Y	LDA (Oper), Y	B1	2	5*

* Add 1 if page boundary is crossed.

LDX

LDX *Load index X with memory* LDX

Operation: M → X

N Z C I D V

✓ ✓ - - - -

(Ref: 7.0)

Addressing Mode	Assembly Language Form	OP CODE	No. Bytes	No. Cycles
Immediate	LDX # Oper	A2	2	2
Zero Page	LDX Oper	A6	2	3
Zero Page, Y	LDX Oper, Y	B6	2	4
Absolute	LDX Oper	AE	3	4
Absolute, Y	LDX Oper, Y	BE	3	4*

* Add 1 when page boundary is crossed.

LDY

LDY *Load index Y with memory* LDY

Operation: M → Y

N Z C I D V

✓ ✓ - - - -

(Ref: 7.1)

Addressing Mode	Assembly Language Form	OP CODE	No. Bytes	No. Cycles
Immediate	LDY #Oper	AØ	2	2
Zero Page	LDY Oper	A4	2	3
Zero Page, X	LDY Oper, X	B4	2	4
Absolute	LDY Oper	AC	3	4
Absolute, X	LDY Oper, X	BC	3	4*

* Add 1 when page boundary is crossed.

LSR

LSR *Shift right one bit (memory or accumulator)*

LSR

Operation: Ø → [7|6|5|4|3|2|1|0] → C

N Z C I D V

Ø ✓ ✓ — — —

(Ref: 10.1)

Addressing Mode	Assembly Language Form	OP CODE	No. Bytes	No. Cycles
Accumulator	LSR A	4A	1	2
Zero Page	LSR Oper	46	2	5
Zero Page, X	LSR Oper, X	56	2	6
Absolute	LSR Oper	4E	3	6
Absolute, X	LSR Oper, X	5E	3	7

NOP

NOP *No operation*

NOP

Operation: No Operation (2 cycles)

N Z C I D V

— — — — — —

Addressing Mode	Assembly Language Form	OP CODE	No. Bytes	No. Cycles
Implied	NOP	EA	1	2

ORA

ORA *"OR" memory with accumulator*

Operation: A V M → A

N Z C I D V
✓ ✓ — — — —

(Ref: 2.2.3.1)

Addressing Mode	Assembly Language Form	OP CODE	No. Bytes	No. Cycles
Immediate	ORA #Oper	Ø9	2	2
Zero Page	ORA Oper	Ø5	2	3
Zero Page, X	ORA Oper, X	15	2	4
Absolute	ORA Oper	ØD	3	4
Absolute, X	ORA Oper, X	1D	3	4*
Absolute, Y	ORA Oper, Y	19	3	4*
(Indirect, X)	ORA (Oper, X)	Ø1	2	6
(Indirect), Y	ORA (Oper), Y	11	2	5

* Add 1 on page crossing

PHA

PHA *Push accumulator on stack*

Operation: A ↓

N Z C I D V
— — — — — —

(Ref: 8.5)

Addressing Mode	Assembly Language Form	OP CODE	No. Bytes	No. Cycles
Implied	PHA	48	1	3

PHP

PHP *Push processor status on stack*

Operation: P↓

N Z C I D V
— — — — — —

(Ref: 8.11)

Addressing Mode	Assembly Language Form	OP CODE	No. Bytes	No. Cycles
Implied	PHP	Ø8	1	3

PLA

PLA *Pull accumulator from stack*

Operation: A ↑

N Z C I D V
✓ ✓ – – – –

(Ref: 8.6)

Addressing Mode	Assembly Language Form	OP CODE	No. Bytes	No. Cycles
Implied	PLA	68	1	4

PLP

PLP *Pull processor status from stack*

Operation: P ↑

N Z C I D V
From Stack

(Ref: 8.12)

Addressing Mode	Assembly Language Form	OP CODE	No. Bytes	No. Cycles
Implied	PLP	28	1	4

ROL

ROL *Rotate one bit left (memory or accumulator)*

Operation: M or A: [7|6|5|4|3|2|1|0] ← [C] ←

N Z C I D V
✓ ✓ ✓ – – –

(Ref: 10.3)

Addressing Mode	Assembly Language Form	OP CODE	No. Bytes	No. Cycles
Accumulator	ROL A	2A	1	2
Zero Page	ROL Oper	26	2	5
Zero Page, X	ROL Oper, X	36	2	6
Absolute	ROL Oper	2E	3	6
Absolute, X	ROL Oper, X	3E	3	7

ROR

ROR *Rotate one bit right (memory or accumulator)*

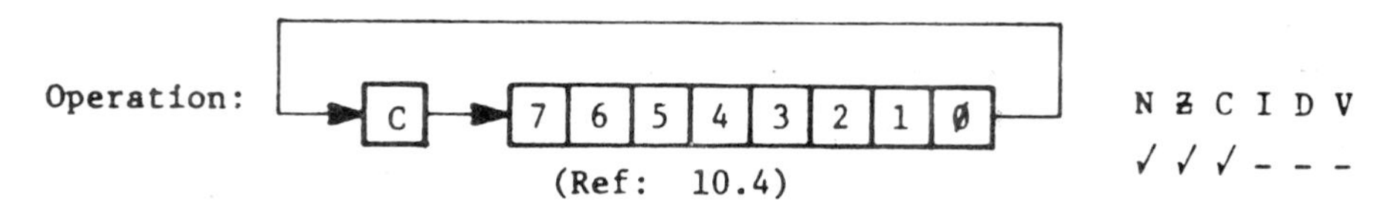

(Ref: 10.4)

Addressing Mode	Assembly Language Form	OP CODE	No. Bytes	No. Cycles
Accumulator	ROR A	6A	1	2
Zero Page	ROR Oper	66	2	5
Zero Page,X	ROR Oper,X	76	2	6
Absolute	ROR Oper	6E	3	6
Absolute,X	ROR Oper,X	7E	3	7

Note: ROR instruction will be available on MCS650X micro-processors after June, 1976.

RTI

RTI *Return from interrupt*

Operation: P↑ PC↑

N Ƶ C I D V

From Stack

(Ref: 9.6)

Addressing Mode	Assembly Language Form	OP CODE	No. Bytes	No. Cycles
Implied	RTI	4Ø	1	6

RTS

RTS *Return from subroutine*

Operation: PC↑, PC + 1 → PC

N Ƶ C I D V

\- - - - - -

(Ref: 8.2)

Addressing Mode	Assembly Language Form	OP CODE	No. Bytes	No. Cycles
Implied	RTS	6Ø	1	6

SBC

SBC *Subtract memory from accumulator with borrow* **SBC**

Operation: A - M - $\overline{C}$ → A

Note: $\overline{C}$ = Borrow (Ref: 2.2.2)

N Z C I D V
✓ ✓ ✓ — — ✓

Addressing Mode	Assembly Language Form	OP CODE	No. Bytes	No. Cycles
Immediate	SBC #Oper	E9	2	2
Zero Page	SBC Oper	E5	2	3
Zero Page, X	SBC Oper, X	F5	2	4
Absolute	SBC Oper	ED	3	4
Absolute, X	SBC Oper, X	FD	3	4*
Absolute, Y	SBC Oper, Y	F9	3	4*
(Indirect, X)	SBC (Oper, X)	E1	2	6
(Indirect), Y	SBC (Oper), Y	F1	2	5*

* Add 1 when page boundary is crossed.

SEC

SEC *Set carry flag* **SEC**

Operation: 1 → C

(Ref: 3.0.1)

N Z C I D V
— — 1 — — —

Addressing Mode	Assembly Language Form	OP CODE	No. Bytes	No. Cycles
Implied	SEC	38	1	2

SED

SED *Set decimal mode* **SED**

Operation: 1 → D

N Z C I D V
— — — — 1 —

(Ref: 3.3.1)

Addressing Mode	Assembly Language Form	OP CODE	No. Bytes	No. Cycles
Implied	SED	F8	1	2

SEI

SEI *Set interrupt disable status*

Operation: 1 → I

N Z C I D V
— — — 1 — —

(Ref: 3.2.1)

Addressing Mode	Assembly Language Form	OP CODE	No. Bytes	No. Cycles
Implied	SEI	78	1	2

STA

STA *Store accumulator in memory*

Operation: A → M

N Z C I D V
— — — — — —

(Ref: 2.1.2)

Addressing Mode	Assembly Language Form	OP CODE	No. Bytes	No. Cycles
Zero Page	STA Oper	85	2	3
Zero Page, X	STA Oper, X	95	2	4
Absolute	STA Oper	8D	3	4
Absolute, X	STA Oper, X	9D	3	5
Absolute, Y	STA Oper, Y	99	3	5
(Indirect, X)	STA (Oper, X)	81	2	6
(Indirect), Y	STA (Oper), Y	91	2	6

STX

STX *Store index X in memory*

Operation: X → M

N Z C I D V
— — — — — —

(Ref: 7.2)

Addressing Mode	Assembly Language Form	OP CODE	No. Bytes	No. Cycles
Zero Page	STX Oper	86	2	3
Zero Page, Y	STX Oper, Y	96	2	4
Absolute	STX Oper	8E	3	4

STY

STY Store index Y in memory

Operation: Y → M

N Z C I D V

\- - - - - -

(Ref: 7.3)

Addressing Mode	Assembly Language Form	OP CODE	No. Bytes	No. Cycles
Zero Page	STY Oper	84	2	3
Zero Page, X	STY Oper, X	94	2	4
Absolute	STY Oper	8C	3	4

TAX

TAX *Transfer accumulator to index X*

Operation: A → X

N Z C I D V

✓ ✓ - - - -

(Ref: 7.11)

Addressing Mode	Assembly Language Form	OP CODE	No. Bytes	No. Cycles
Implied	TAX	AA	1	2

TAY

TAY *Transfer accumulator to index Y*

Operation: A → Y

N Z C I D V

✓ ✓ - - - -

(Ref: 7.13)

Addressing Mode	Assembly Language Form	OP CODE	No. Bytes	No. Cycles
Implied	TAY	A8	1	2

TYA — TYA *Transfer index Y to accumulator* — TYA

Operation: Y → A

N Z C I D V
✓ ✓ — — — —

(Ref: 7.14)

Addressing Mode	Assembly Language Form	OP CODE	No. Bytes	No. Cycles
Implied	TYA	98	1	2

TSX — TSX *Transfer stack pointer to index X* — TSX

Operation: S → X

N Z C I D V
✓ ✓ — — — —

(Ref: 8.9)

Addressing Mode	Assembly Language Form	OP CODE	No. Bytes	No. Cycles
Implied	TSX	BA	1	2

TXA — TXA *Transfer index X to accumulator* — TXA

Operation: X → A

N Z C I D V
✓ ✓ — — — —

(Ref: 7.12)

Addressing Mode	Assembly Language Form	OP CODE	No. Bytes	No. Cycles
Implied	TXA	8A	1	2

TXS

TXS *Transfer index X to stack pointer*

TXS

Operation: X → S

N Z C I D V

_ _ _ _ _ _

(Ref: 8.8)

Addressing Mode	Assembly Language Form	OP CODE	No. Bytes	No. Cycles
Implied	TXS	9A	1	2

Appendix C: Solutions to All Exercises

APPENDIX C

Solutions to All Exercises

Unit 2

2-1. Three, 1 in 500 (assuming no drift).

2-2. Nine, since 2 to the 9th power = 512.

2-3. False
- sensitivity threshold
- calibration
- hysteresis
- noise level
- digital step error (roundoff)
- human reading error, such as parallax

2-4.

0000	1000
0001	1001
0010	1010
0011	1011
0100	1100
0101	1101
00	1110
0111	1111

The binary equivalent of decimal 15 is 1111.

2-5. a) .0001100110011001
b) .0999 (since 6553/65536 = .09999084
c) No, see b), above, for the reason why not.

2-6. Since "Gilgamesh" evidently counted in base 6 or 12, he may have counted on the six fingers of each hand!

2-7. a) The "Martians" must have used a base-3 system with the following table of numbers and decimal equivalents.

000	0
001	1
002	2
010	3
011	4
012	5
020	6
021	7
022	8
100	9

b) Base-3 (ternary) computers would be less reliable than our binary machines because of the greater difficulty in distinguishing between three states rather than two, in the presence of noise, etc. A possible advantage of a ternary computer would be greater memory capacity for a given space and cost, assuming that the hypothetical ternary element was the same size and cost as our binary memory cells.

Unit 3

3-1. 1776 (octal)

3-2. $A1 = 10*16+1* = 160+1 = 161 (decimal, via weights)
$A1 = 1010 0001 = 128+32+1 = 161 (decimal, via binary)
$A13F = 41279 (decimal, by either method)
You may find the weight method quicker for the four-digit number.

3-3. a) The largest is 0111 1111 or +127
The smallest is 1000 0000 or −128
b) −128 is 1000 0000

3-4. a) 37 = 0011 0111 BCD
42 = 0100 0010
99 = 1001 1001
b) 0100 0111 = 47; 1010 0011 is not valid (1010 is not BCD)

Unit 4

4-1. 000
001
010
011
100
101
110
111

4-2. 1100 A
1010 B
————
0000 FALSE (CONTRADICTION)
0001 NOT OR (NOR)
0010
0011
0100
0101
0110 S1/2 = X−OR
0111 NOT AND (NAND)
1000 C1/2 = AND
1001
1010
1011
1100
1101
1110 OR
1111 TRUE (EXCLUDED MIDDLE)

4-3.

A	B	OR	NOT-OR
0	0	0	1
0	1	1	0
1	0	1	0
1	1	1	0

4-4.

A	B	NOT-A	NOT-B	C1	C2	D
0	0	1	1	0	0	0
0	1	1	0	0	1	1
1	0	0	1	1	0	1
1	1	0	0	0	0	0

4-5. 29 = 011101
25 = 011001

----

54 110110 Sum
11001 Carry

4-6. 45 = 00101101
145 = 10010001

----

190 = 10111110 Sum
0000001 Carry

4-7. 255 = 11111111
2 = 00000010

? 100000001
The apparant answer is 1. It is wrong because the 8-bit adder overflowed (look at the "9th bit!")

4-8.

A	B	C	S1/2	C1/2	S	C'	C1
0	0	0	0	0	0	0	0
1	0	0	1	0	1	0	0
0	1	0	1	0	1	0	0
1	1	0	0	1	0	0	1
0	0	1	0	0	1	0	0
1	0	1	1	0	0	1	1
0	1	1	1	0	0	1	1
1	1	1	0	1	1	0	1

This is a full adder because its outputs (S, C1) for all possible inputs A,B,C correspond to the outputs of the Full Adder Truth Table (see Table 4-7).

4-9. 1111 1110 −2 (two's complement)
0000 0001 Recomplement (one's complement)
1

0000 0001 two's complement = 2

The magnitude of the result of Example 4-7 is 2. The sign (−) must be remembered after the recomplement.

4-10. The apparent result is 127, but it is incorrect because of the overflow (the sign bit changed from 1 to 0).

Unit 5

5-1. 16

5-2. 2^{10} or 1024

5-3. 23. 11 address, 8 data, R/W, CE, Vcc, and Vss.
Actually, 28 pins are used so as to be compatible with EPROMs.

5-4. a–1, b–4, c–3, d–2

5-5. All MOS memories are random access. R/WM is a more exact terminology.

5-6. EPROM is the most flexible memory for an experimental device, since the code can be erased and altered if a "bug" is found. RAM does not fit the specifications because it isn't permanent.

5-7. When high-speed random access is required.

5-8. A computer that shares the same random access memory space between programs and data. This allows program branching and consequently a flexible response to incoming data (branching).

Unit 6

6-1. a) 4-bit, Intel 4004 or MCS-4
b) 8-bit, Z80, 6502, 8080, and others
c) 16-bit, Intel 8086, TI 99000, MC68000, others

6-2. Yes. There must be more address lines to address more memory. (23 lines in the MC68000 for 8 megawords of memory.)

6-3. 65,536 (2^{16}) assuming 2 bytes per op code.

6-4. Two address bytes (low and high bytes) are required. Two fetches are needed to put the address low and high bytes on the address bus, requiring twice the time of one fetch.

6-5. CMOS is a good choice for many process control applications requiring low power. For example, a sensor or transducer in a CPI field location could be powered by a long-lived battery to eliminate the hazard and complexity of local power sources using higher voltage or transmitting low-voltage power on lines.

6-6. The 4-pin DIP package satisfies these needs:
a) address, data, power, clock, I/O = 27 pins minimum
b) standard package
c) interchangeability due to standard sockets.

6-7. No, not without severely restricting address space. Use of the 23 address lines in the MC68000 would exceed the 40-pin limit.

6-8. The best place to cut down would be address space as it is not needed for these applications.

6-9. The accumulator register is the main interface with memory.

6-10. The program counter. It holds the address of the next instruction, distinguishing it from data.

6-11. No. None.

6-12. Yes, else you cannot program the MPU.

6-13. Instruction fetch and instruction execute phases.

6-14. It depends on the instruction. At least two cycles are needed for fetch and execute, but a code may need as many as six cycles or more if the operations are complex.

6-15. 16-bit wide registers facilitate precision arithmetic.

6-16. a) Move data by incrementing a base address.
b) Timing
c) Counting loops.

6-17. a) The automatic tray dispenser in some cafeterias.
b) A stack of order forms on a spindle or desk tray.
c) Stock in a LIFO stockroom.

6-18. No parentheses are used, reducing memory requirements.

6-19. The stack pointer is used to store and retrieve addresses when subroutines are entered and returne from. It keeps track of the correct address even though subroutines are nested.

6-20. 0111 1001 $79
0001 0100 $14 (Binary Add)

———————————

1000 1101 $8D (Not a legal BCD code)

6-21. It can't.

6-22. To implement conditional program branching.

6-23. Rotate passes bits through Carry; bits are recirculated. Shift is into Carry. Bits are lost after leaving Carry.

6-24. ASL saves bit 7 (sign bit) in Carry
ROR restores from Carry to bit 7

Unit 7

7-1. 3.1662 in 5 iterations

7-2. The recipe is an algorithm but not a computer program since it does not use a language available to any computer.

7-3.

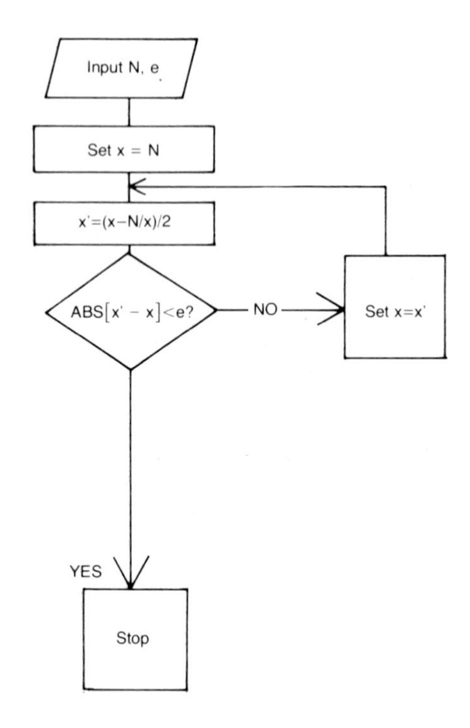

7-4. a) Assembly language
b) Pascal or FORTRAN
c) Assembly or FORTH
d) Learn BASIC

7-5. Larger address space and word size will encourage the use of high-level languages. 32-bit microprocessors are being built with Ada (similar to Pascal) as their "native language."

7-6. STA 1500 (or STA $5DC)
85 DC 05
1000 0101 1101 1100 0000 0101

7-7. ER = PV − SP
This would take about 10 lines of assembly code.

7-8. Since low byte is the only byte needed in zero page it must come first if space is to be saved when using zero page.

7-9. a) LDA #%01000000
b) LDA #$40
c) A9 40 or actually 1010 1001 0100 0000

7-10. a) LDA #09 or A9 09
b) LDA *FE or A5 FE
c) LDA A5 40 or AD 40 A5
d) The penalty for using LDA FE 00 instead of LDA *FE is one byte and one clock cycle.

7-11. Since the data to store is already in the accumulator, it is not sensible to give it again in the next byte, as in the immediate mode of LDA.

Unit 8

8-1. a) 100
b) START
c) $500-$502
d) $5001
e) It disregards them during assembly.

8-2. a) 10 LDA $2010 ;GET DATA FROM $2010
20 STA $1000 ;MOVE TO $1000
30 BRK
b) 0500 AD 10 20 8D 00 10 00 00, 7A
017A

8-3. a) $29 incorrect
b) Test Carry flag (should be cleared first)

8-4. a) Source

100 CLC	2 (CLOCK CYCLES)
110 LDA $500	4
120 ADC $400	4
130 STA $600	4 (14 TOTAL)

Object
200 18
201 AD
202 00
203 05
204 6D
205 00
206 04
207 8D
208 00
209 06
210 00
211 00
b) Clock cycles 14 istead of 13

8-5. 500 A5
501 10
502 18
503 65
504 12
505 85
506 14
507 A5
508 11
509 65
510 13
511 85
512 15
513 00
b) Add another LDA, ADC, STA, before BRK
c) No. Yes. Must be cleared before adding.

8-6. C8 = 200 = 1 ms
FF = 255 = 1.275 ms
255*1.275 = 325 ms (ans.)

8-7. Substitute RTS for BRK. (It would be a good idea to save the status registers, accumulator, and index registers by pushing them on the stack and restoring them on return from the subroutine.)

Unit 9

9-1. 2

9-2. a) Line 30 LET X=66
b) No effect if X=65, 6 is printed
If X=77, nothing would be printed (not 0).

9-3.
```
10 LET N=1
20 PRINT N*N
30 N=N+1
40 IF N=10 GOTO 60
50 GOTO 20
60 END
```

9-4. It will print TRUE and then print FALSE on the next line.

9-5. 4.4, 18.0, −21, 30

9-6.
```
10 FOR N=2 TO 10 STEP 2
20 PRINT N*N*N
30 NEXT N
```

Unit 10

10-1. RADIUS @ 2 * PI 10000 */ . (ret.) 31

10-2. 1 3 5 5 <TOP
1 3 5
3 5 1 1
1 5
1 3 5

10-3. 15, 15, 4, 0, 20, 2, 12

10-4. a) 12 CONSTANT DOZEN
b)
```
5 CONSTANT WIDTH
VARIABLE DEPTH
5 DEPTH !
WIDTH DEPTH @ * . 25
10 DEPTH !
WIDTH DEPTH @ * . 50
15 DEPTH !
WIDTH DEPTH @ * . 75
```

10-5. No, because the intermediate result of (*/) is 785400. This result fits into the 32-bit double precision product of "times-divide" but would overflow the 16-bit product that would be formed from two separate operations * and /. (Recall that the maximum signed 16-bit number is 32767.)

10-6.
```
: COUNT 1 BEGIN DUP . DUP 10 = NOT
        WHILE 1+
        REPEAT ;
```

10-7.
```
: TEST DUP 5 > IF ." GREATER"
          ELSE DUP 5 = IF ." EQUAL" EXIT
                  THEN ." SMALLER"
          THEN ;
```

If you got this one, congratulations! You are now a FORTH programmer.

10-8. The program (DO LOOP) would ignore the last item on the stack.

Unit 11

11-1. If the sampling rate of the recording system is twice the highest frequency in the audio input there is no loss of fidelity in playback, according to Shannon's Theorem. The digital recording could be better because it is more stable with time (no grooves or tape to wear out).

If the highest response of the human ear is 20,000 Hz, the sampling rate should be at least 40,000/sec and perhaps as high as 80,000 or 100,000.

11-2. If T[s] is small compared with T[1], the usual case, then at each step the output D[n] changes only a small fraction of the remaining difference (error). As T[s] increases the amplitude of changes in D[n] increases proportionally. In the limit, if T[s] was made very much larger than T[1], nearly the entire gap would be closed on the first step.

11-3. a) Could be part of a more complex algorithm.
b) The digital form is more stable and can simulate larger time constants than practical with "real" components.
c) Component space and cost are saved if function is time-shared.

11-4. The difference equation was already in correct form. The output V[n] has only present and past values of error in the RH side.

11-5. Line 5070

11-6. Line 5060 filters the "raw error" E2 through a first-order filter time constant TF and the sampling time TS. The PID, line 5070, operates on the filtered error input F2.

11-7. The error inputs are differences rather than absolute values as in Eq.(11-6). (F2-F1) is the present error increment (current minus previous value of the filtered error) and (F2-F1)−(F1-F0) represents the current less the previous difference. It is used in the same way as (E[N]−E[N−1]) of Eq. (11-6) but yields a differential valve output rather than an absolute position.

11-8. Lines 5080 and 5090 exchange and update the current and past value of filtered error, so that when the algorithm is entered the next time they will have the correct values.

Index

Index